MANCHESTER PHYSICS

1641–1870

Volume 1

in the series
Three Centuries of
Manchester Physics

Robin Marshall

First published in paperback in 2017 by Champagne Cat

British Library Cataloguing in Publication Data.
A catalogue record for this book will become available from
The British Library.

ISBN 978-1-9732609-5-0

A CHAMPAGNE CAT PRODUCTION

Contents

List of Illustrations

Preface

About the Series

This preface spans all the volumes of this series of books covering the three centuries (and a few decades) of Manchester Physics. As well as providing generality, it also offers a glimpse of the content of other volumes.

I have tried to write this book from the perspective of a scientist interested in history rather than a historian interested in science. I beg the indulgence of historians, whose rules I might not obey as scrupulously as they do. In my defence I would argue that I have striven to focus on fact and to minimise opinions and judgements and before that is taken as a judgement on historians, I recognise their right to interpret, and assert my right to accept or not. I confess that I am unable to read Macaulay's 'Whig interpretation of history' without a sense of outrage at some of his biased descriptions and would simply point to his account of the siege of Derry to anyone who disagrees with me. And having said all that, and reviewed what I have written, of course I have fallen into the temptation to make judgements, but not often.

In trying to get a feel for the mood of the people in the 17th century when a university in the village of Manchester was first mooted, I chose to read contemporaries as much as possible. There are a variety of first hand accounts of much of what happened in Manchester, setting up the eventual college that became a university and how physics developed as a research and teaching enterprise. I can rely on these and summarise the main sources as follows:

Alderman, historian and lifetime member of the Owens College governors, Joseph Thompson, was commissioned by the College to write a history of its foundation and growth [1] and his 671 page volume is a wealth of detail of procedures, conversations, minutes, all recorded with a minimal overlay of opinion. He was also asked to write a history of the foundation of the Lancashire Independent College in Whalley Range [2], which provides general insight into an alternative seat of learning. His books contain informative drawings and superb illustrations.

The book *Manchester Old and New* [3] by William Arthur Shaw, Fellow of Owens College is a delight and is filled with reproductions of Manchester scenes painted by Henry Edward Tidmarsh.

Philip Hartog, a former chemistry student and lecturer at Owens, who eventually became a prominent educationalist, wrote a factual history [4] of the first half century of Owens and this also contains much detail, especially about the various departments in 1901. It also contains some fine photographs.

An appropriately jubilant and happy account [5] of the Owens jubilee celebrations of 1901 was produced by Miss Josephine Laidler, who arranged and edited the text together with Mr A R Skemp who saw the 'little book' of 224 pages through to the press. Within its interestingly broad folio are extracts from the Manchester Guardian and a number of contemporary photographs.

A record of the Physical Laboratories was produced in 1906 to commemorate the 25th anniversary of the election of Arthur Schuster to a professorship in the Owens College. This book [6] contains a list of all physics alumni and their publications whilst at Manchester. There are also many photographs of the laboratories and lecture rooms. I found a battered copy of this book in the basement of the physics building, had it restored and placed it for safe keeping in the University Library. Schuster wrote a fragmented biography of himself and others in 1932 [7]. His memory in this book can be shown to be imperfect and wherever there was a variance, I preferred to use what he wrote at the time to what he wrote over half a century later.

The *Letter Book*, maintained in the Physics Department from 1873 to 1882 provides an enormously detailed insight into what the laboratory was doing during that period. It is reproduced in full in Chapter 3, which is contained in Volume 2 and it is not an indulgence; it shows the history of the physics department at the time, better than a historian can reproduce it.

From the date he finally took up his appointment in 1872, Balfour Stewart kept detailed log books of the experiments carried out in the laboratories, which were used by students and staff alike. These are an equally fascinating account of the work, the people who did the work and the conditions under which they worked. Extracts are also reproduced in Chapter 3 in Volume 2. The complete log books are greater than the volume of this book and contain sheaves of recorded numbers. They are available for the dedicated in the Physics Archive in the University Library.

Chemistry professor Henry Enfield Roscoe, who taught some of the physics courses in the 1850s, (in the absence of a dedicated physics professor) wrote an autobiography [8] which covers all of his time at Owens College and his observations on its early history.

For a graphic portrayal of what it was like to live in Manchester through the ages, the *Court Leet Records* [9] of the Manor of Manchester from 1552 up to 1846, when the newly formed Corporation bought the baronial rights off Sir Oswald Mosely, provide a developing insight into the matters of importance to the town, designed to ensure a smooth life for some of its inhabitants.

Henry Bright's historical sketch of the Warrington Academy [10] is a useful reference to that establishment.

William Axon's *The Annals of Manchester* [11] is a handy reference for dates.

The largely philosophical commemorative book *Portrait of a University, 1851–1951* [12] by Henry Buckley Charlton, has not aged well and offers nothing more than some references to other more useful sources. There is an absence of physics in this book by the admitted erudite poetry theorist, apart from some of Schuster's internal university politics. Nobel prizes in physics cut no ice with the arts-orientated Charlton.

A series of essays edited by D S L Cardwell, tracing the founding of the Manchester Mechanics' Institution in 1824 to its establishment as a university, UMIST, can be found in [13]. The collection of Cardwell's papers on the development of science and technology in nineteenth century Britain, with the sub-title *The Importance of Manchester* [14] was indispensable and his biography of Joule [15] is a *tour de force*.

A day of 'Bragg Centenary' lectures was held on the 21st of March 1990 in Manchester and the video tape of the proceedings has been used to provide quoted extracts from Nobel Laureate Sir Nevill Francis Mott, Manchester graduate Cecil Arthur Beevers and crystallographer Henry Lipson, all of whom were staff members or researchers in Bragg's department between 1919 and 1937.

George Rochester, co-discoverer of strange particles and *de facto* of the strange quark in 1947, gave a lecture in Manchester about the discovery in 1985. The lecture was recorded in audio and video and Rochester's personal remarks form the basis of the textual content in this book about the discovery. Likewise, Charles Clifford Butler, the other co-discoverer recorded his memories on video tape in 1997, the 50th anniversary of the discovery. His factual comments also provide content here.

Acknowledgements

For information, photographs, corrections, discussions and/or correspondence I am utterly indebted to (A–Z):

Epiphany Appleseed[1], J R J 'Roger' Bennett, David Binnie, Graham Bowden, Douglas Broadbent[2], Sir Clifford C Butler, Ian Butterworth, John Campbell[3], Noel Corcoran, R L A 'Les' Cottrell, George Crossley, Sir Sam Edwards, Lord Flowers[4], Peter Ford, Alan Gall, Bernie Geiger (Grandson of Hans), John Gilmore (great grandson of Arthur Schuster), Dehn Gilmore (great-great- granddaughter of Arthur Schuster), Alison Goddard, Robert Hanbury Brown, Leslie Alan Hodson, Alan Hughes, Frank James, John James (Son of R W James), Jeremy Kahn (son of Franz Kahn), Rajesh Kochhar, Sir Bernard Lovell, Malcolm McCausland, Paul G Murphy, James Peters, J E B Ponsonby, Brian Pollard, Arthur Porter, Brian Sullivan, James Sumner, Derek Shaw, B S Shylaja, Brian Thompson, Patience Thomson (daughter of W L Bragg and wife of the grandson of J J Thomson)[5], Alan Watson, Raj Kumar Jones[6], Emil Wolf, Sir Arnold Wolfendale and Peter Zernik.

Images and Illustrations

The illustrations are a strong feature of this book, especially the presentation of a specific kind of image art used by the author. This art is described in his publication [17] in the *Journal of the Perthshire Society of Natural Science*. The first professor at Owens College, Archibald Sandeman, endowed the library in Perth and was the second attempt at colour processing of a historic image by the author. The first attempt was

---

[1] I am especially in debt to Epiphany Appleseed for reading, correcting and criticising the whole manuscript in finest detail and tactfully encouraging its final production in the form I strove to write it.

[2] I received lectures from Douglas in my first year as an undergraduate. He looked after Einstein's signature, written in chalk on a Manchester blackboard for 25 years and then identified all graduates in the 1942 photograph which appears in Volume 3.

[3] For anyone seeking a detail on Rutherford, or its confirmation, John is the first person, and probably the only one, to ask.

[4] In his nuclear physics course, Brian Flowers taught me three things: There are $\pi \times 10^7$ seconds in a Flowers year, the neutron lifetime is a quarter of a Flowers hour and no lecture should exceed a Flowers microcentury.

[5] Having agreed to a date – tea at the Ritz, to discuss newly found WW1 papers concerning her father, which form the core of my book *'Physicists at War'* [16], Patience stood me up in favour of minor royalty. I forgive her.

[6] Raj is to Schuster (and Dalton) as John Campbell is to Rutherford.

the driving force. The splendid painting of Joule in his prime by George Patten, created in 1864, was destroyed during the 1940 Manchester Blitz. Somewhat as a consolation, a high quality engraving of the painting had been made in 1892 and used as a frontispiece to the biography of Joule that was written by Osborne Reynolds and published in the *Memoirs of Manchester Literary and Philosophical Society* [18]. It is an interesting feature of line engravings that the rather arbitrary line grid can be removed using a notch filter, almost as well as with a half tone. A study of other (surviving) paintings by the artist of other sitters revealed that Patten often used the same clothes, which the sitter might never even have worn. The chair in which Joule was sitting appears regularly and the room wallpaper and drapery were not unique. All those colours could therefore be retrieved. All that remained was Joule's apparatus and his features. The apparatus has survived and is displayed by the Science Museum in Kensington, this yielding more colours. Finally, the core element of the colour processing was applied to the features, a classic mathematical inverse problem where a 256 dimensional vector (grey-scale) is converted to a 3×256 matrix (RGB, with 8 bits each colour). This process uses the usual Tikhanov regularisation (to give the most plausible solution) together with the constraint of the Kuhn-Tucker algorithm (the final image must be positive and not have negative components). The result for Joule can be seen in Figure 1.32 on page 82 in Chapter 2 of this Volume.

Colour processed images are the copyright of the author and are marked as such in the caption. They are available free for personal use from the publishers (see above for details) and generously for commercial or display purposes.

Historic images in the public domain are assigned, where necessary, in the figure captions. Those that have been generously and courteously provided for use here are acknowledged as such, specifically in the caption.

Departmental and University archive images are denoted accordingly. In many cases, the images in the form of prints, negatives or glass plates were rescued from the Manchester Physics Department basement during a refurbishment and deposited in the University John Rylands Library.

The author has also used images from his own personal archive, which consists of his own photographs and those freely given during the course of writing this book.

The Five Volumes of this Book – Highlights

Volume 1 is this one. See the index for contents.

Volume 2 continues the story from 1870, when the professional physicist-astronomer Balfour Stewart was appointed as professor of physics to succeed the mathematical-administrative orientated William Jack. He was the first holder of the new Langworthy Chair and his title included the word 'Physics', instead of 'Natural Philosophy'. Arthur Schuster emerged during this period. He was appointed to a new chair of mathematical physics in 1881 and then, as the second Langworthy Professor, handed over to Ernest Rutherford in 1907, when this volume ends. A transcript of the Department's 'Letter Book' from 1871 to 1881, correspondence mainly with instrument makers, provides a window on daily life.

Volume 3 covers the first period of the Nobel era from 1907 to 1937. The Physical Laboratories of the University dominated physics not in Manchester but in the whole world, with Ernest Rutherford and then W L Bragg at the helm. This volume is filled with nuclear physics, the discovery of the atomic nucleus, the transmutation of elements, all during Rutherford's reign, followed by crystallography guided by Bragg. Physics at the Tech slowly gains a toe-hold. Einstein's first lecture in Britain, held in the Whitworth Hall in 1921 is given extensive coverage.

Volume 4 covers the final period of this book, from 1937 onwards. P M S Blackett, appointed just before WW2, was the fifth Langworthy Professor and the third Nobel prize-winner to be Director of the Physical Laboratories. He was followed by Samuel Devons, who can be credited with diversification during his brief tenure, but who is otherwise forgotten like a bad dream. This volume covers the good and the bad of the post-war era – the discovery of V-particles in Coupland Street, which changed physics for ever – and the University's poor approach to post-war expansion, which was noted, arrested and re-formulated. The volume ends in 1967, when the splendid new physics building was occupied.

Volume 5 collates every physics student in Owens College from the start of records under Balfour Stewart up to 1951, when the list became long. There are short micro-biographies of interesting characters. This volume, the e-book of which will be free to purchasers of any of the above four volumes, also includes a photo album derived from various archives, showing Manchester physics and physicists as never before seen.

About the Author

Most of the author's career was spent as a particle physicist, working successively at the University of Manchester, the German Electron Synchrotron (DESY), MIT, the Daresbury Laboratory, and the Rutherford Appleton Laboratory, before returning to the University of Manchester as professor in 1992.

For his work in particle physics, especially measurements of the weak isospin of the b-quark, which showed that the t-quark must exist, twenty years before its discovery, he was elected as a Fellow of the Royal Society and awarded the Max Born Medal and Prize.

As well as physics, he also secured funding to apply neural networks to cancer management. For three years, after relinquishing the leadership of the particle physics group in Manchester, he was a research professor in physics and life sciences and studied the crystallography of ice.

In researching the five volumes of this book, *Three Centuries of Manchester Physics*, he read many hundreds of papers, some of them hundreds of years old. This was to ensure that he understood what people of interest actually said, and not what others said they had said. He also read the relevant parts of hundreds of books, to trace the story back to source. Over the years he has built up a collection of historical photographs, some never seen in public before. These form a crucial part of this book, together with his own individual style of art, re-colouring naturally, historic images originally shot in monochrome.

He now lives in the South of France, near Avignon, painting, writing and running the publishing company *Champagne Cat*. As well as writing this book, he also dealt with its typesetting and jacket design, before managing the publishing aspects.

Prof Robin Marshall FRS FInstP

7

Chapter 1
1641–1851

1641: Petitions for a University in Manchester

Physics in Manchester has not always been confined within the premises of what is now the University of Manchester. Indeed, the first teaching of physics as a subject in its own right up to university entrance level took place in an independent dissenting college and Joule carried out his research from home, declining the offer to be a professor at Manchester.

Yet physics in Manchester became increasingly focussed on the University with time and so this chapter will trace the evolution and establishment of a university in the city. Although the old universities of England and Scotland were an integral part of the church and theological infrastructure of the two countries, educating and preparing future priests, they also housed the most eminent mathematicians and natural philosophers. Of especial note was the Merton College school of natural science, including the *Doctor Profundus* by Thomas Bradwardine (ca 1290–1349). The main kinematic properties of uniformly accelerated bodies, still erroneously assigned to Galileo by many, were derived by the Merton scholars in the early 14th century. Manchester had a lot of catching up to do.

New universities emerged in the 19th century with the intention to teach more than theology, or even to exclude it altogether; Manchester and University College London being prime examples. The inclusion of some science within a mainly theological curriculum was not unknown in the past. It was widely held, especially by non-conformist colleges of the nineteenth century that future priests should especially be taught astronomy in order to more fully appreciate what was held to be the wonder of their God's creation.

Although Manchester was eventually established as an ostensibly secular university in 1880, this would not have been the case had earlier efforts been successful. The first recorded instance of an attempt to establish a university in the North of England appears to have been made when Ripon petitioned King James I in 1603, the first year of his reign. According to J T Fowler [19], wrong arguments were used and it came to nothing. According to me, it would have come to nothing even if the 'right' arguments had been used.

The first recorded occasion when the people of Manchester made representations to Parliament for a university was in 1640. If this had been successful, it would have established a theological university, which is what all universities in the world were at the time. But the attempt failed and with hindsight, it might not have been the most propitious time to make the application.

During the eleven years preceding Manchester's petition, King Charles I had not once called a parliament, whereupon a new one (The Short Parliament) was 'elected' in 1640. It lasted only from the 13th of April to the 5th of May 1640 before it disagreed with Charles and was dissolved. The so-called Long Parliament was formed on the 3rd of November and because a new Act of Parliament proscribed that it could only be dissolved by the members themselves, it is possible that the group of Manchester men who prepared the petition for a university were lulled into a sense of optimism by the potential longevity of a new kind of parliament. Their optimism would have been shaken had they had access to the accounts of proceedings of the parliament as eventually published by John Rushworth in his 884 page book [20] *Historical collections containing the principal matters which happened from the dissolution of the Parliament on the 10th of March* $162\frac{8}{9}$ *until the summoning of another Parliament which met at Westminster, April 13, 1640.*

Rushworth took care to state on the title page that the accounts were impartially related, setting forth only matters of fact in order of time. He also said that he preferred 'not to write in the usual form of historians, to pretend to have seen into the dark closets of States-Men and Church-Mens minds.' These many pages not only give a riveting account of this most momentous period in English history, but show the jockeying for power and control by the King, the Archbishop of Canterbury and the parliamentarians over the universities at Oxford and Cambridge. To expect a new university to be inserted into this maelstrom would have been a naivety, had all the facts been known.

The group of petitioners consisted of the 'Nobility, Gentry, Clergy, Freeholders and other inhabitants of the northern part of England'. One of their members was Henry Fairfax, the Rector of Ashton-under-Lyne, a rare non military member of a military family. The parish church of Ashton-under-Lyne at that time held almost equal strength with the collegiate church of what was then the baronial village of Manchester, or Mamecestre as it was called. It was Henry who sent the petition, together with a personal covering letter to his republican orientated brother Ferdinando, the second Lord Fairfax (see Figure 1.1), who had just been returned as Member for Yorkshire in the Long Parliament.

Henry's letter is dated the 20th of March 1640[7] but due to the phasing of the legal year at the time, a system enduring until 1752 when Britain finally adopted the Gregorian calendar, the date of the 20th of March 1640 actually followed November 1640, so the Long Parliament was sitting at the time of the letter.

Figure 1.1:

Ferdinando Fairfax. The first man on record to demure at the thought of a University in Manchester, thus hindering the advancement of science.

Composition © 2017 Robin Marshall.

It has now almost slipped into oblivion that the year number used to change on the 25th of March each year (Lady Day), so that the day after the 24th of March 1640 was actually the 25th of March 1641. In 1751, it was decided to kill two birds with one stone. 1751 itself was the prime casualty, having only 282 days and the dates from the 1st of January to the

---

[7]For reasons about to be explained, the year 1641 is used in the title of this volume, to maintain consistency with the dating system used in 1870.

24th of March 1751 simply never existed in English Law, a move designed to bring the start of the year into line with the rest of the 'civilised' world.

To deal with the second 'bird', the 2nd day of September 1752 was followed by the 14th, to align with Pope Gregory XIII, a most unpalatable action at the time and the reason why the British were almost the last country to do it. By burying the Gregorian news in a year when the seismic event was the shift in New Years' Day, attention was focussed away from Pope Gregory. One result, in order to placate the taxpaying population who would have received an annual tax bill a mere 353 days after their last one, the tax year was moved to start on the 6th of April, where it has remained ever since. Counting backwards in years, the Manchester petition was sent on the 20th of March 1641 (New Style) and the covering letter, transcribed in the *Fairfax Correspondence* [21] bearing a written date of the 20th of March 1640 (Old Style), read as follows:

> 'May it please your Lordship,
>
> I have here inclosed some propositions lately made at Manchester, in a public meeting there, concerning an university; which, if you please to consider what good it may bring to our whole North, and other parts; what glory to the Parliament to be the founder of that, and what honour to your lordship to be the chief agent in it; posterity may bless you, and the work itself will speak that the like hath not been in England (if Cambridge be the last), not of two thousand years.
>
> Your Lordship's ever faithful and loving brother
and servant,
>
> Henry Fairfax'

Whether or not politician Ferdinando found his brother's presentation, coming from a man of the cloth, rather than a fellow man of steel, to be naive or obsequious he did not show it in his reply. Henry's letter had been accompanied by the petition of considerable length and it is reproduced in full in [21]. Here a summary is enough. It began:

> 'To the right honourable the High Court of Parliament, now assembled, the humble petition of the nobility, gentry, clergy, freeholders, and other inhabitants of the Northern parts of England, Humbly showeth, etc etc.'

There were six main points to the petition, which can be summarised as follows using the original language:

1. Oxford and Cambridge were a great distance from Manchester.

2. There were so few places at Oxford and Cambridge that Manchester's young men were kept out by those with money and ended up with a poorer education and having to go into the church.

3. A university in Manchester would convince and discourage papists by teaching them the proper way.

4. Manchester had local benefactors who would fund poor students.

5. Manchester's honour and learning had been eclipsed by being so far from the Court and universities.

6. Manchester was a fit place and moreover has the convenience of an existing large and ancient college, intended for the purpose by Lord Strange.

Joseph Thompson [1], on page 514 of his book, inexplicably identified the 'large and ancient' college mentioned in the petition as the Grammar School, founded by the Bishop of Exeter, Hugh Oldham. The word 'college' had a strict legal and ecclesiastical meaning in 1640 and Hugh Oldham's school was not one. Moreover, the Grammar School was a small single story building, erected only 135 years before the petition, hardly ancient and indeed, a mere stub near the Eastern side of the 'College of Clergy'. The ancient 'College of Clergy' itself was a spectacular stone edifice, built on the site of the original manor house of the 'presumptuous and fiery' fifth baron of Manchester Robert Grelly on a classic fortified location at the junction of two rivers. Grelly, variously spelt as Greslet, Gresley, Gredle or Grelle, had witnessed the signing of the Magna Carta. The eighth baron Thomas Grelly granted the manor to John la Warre in 1309 and one of John's sons Thomas secured a King's licence to convert the manor into a College of Clergy in 1421, whereupon a major rebuilding was carried out in 1422 for this purpose.

Although there is no impact on the Manchester physics that ultimately transpired, the date of 1421 has been questioned [22] by Hibbert *et. al.* because the charter chest was 'carried away by Colonel Birch and his commonwealth rabble' during the civil war and subsequent historians were unable to resolve the partial paradox that some parts of the buildings looked older than 1421.

Thus it was that a College, in its strictest theological sense, was first established in Manchester in 1421 (or thereabouts). But for the oppressive intervention of self preservationists, it would have become the first college of the ancient University of Manchester.

The Church had enjoyed possession of the building until it was given to the Earl of Derby by Edward VI, who had dissolved the College of Clergy on his accession in 1547. In 1640/1, when Lord Strange (James Stanley, the future 7th Earl of Derby), had devised plans of his own for this former College, he was heir apparent to his father William, the current Earl of Derby but soon became embroiled in other pressing matters. He led the Royalist siege [23] of parliamentarian Manchester during the course of which, the College was variously used as a gunpowder store and prison, materially ending up almost a ruin. Particulars printed by order of Parliament on the 9th of July 1642, concerning 'The beginning of civil warres in England or terrible news from the North' went on to say about Lord Strange that 'he was much perplexed in mind.' The first blood of the civil war had now been shed in Manchester where the siege had begun on the 4th of that month.

The reason why James Strange was much perplexed in mind was because, according to reports [23], he had lost over 200 men against a mere four of the besieged Mancunians, one of whom had died when his comrade's musket had gone off 'unawares'. The House of Commons ordered that the deliverance of Manchester should be observed by all churches and chapels in Lancashire. Then on the 29th of September of that year 1642, William Stanley, the sixth Earl of Derby died at home on the banks of the river Dee near Chester and his son James inherited.

To the best of my knowledge, it has not been noted before that James Stanley, Lord Strange had in fact, laid siege to a modestly sized market village whose largest residence was his own father's second home. With the defeat of the Royalists, the estate of the new earl, along with the former College buildings, was confiscated by the Commonwealth. For his part in the massacre of a large fraction of the population of Bolton, James Strange was beheaded on the 15th of October 1651 outside *Ye Old Man and Scythe* public house.

After the Restoration, the estate was returned to his widow, the renowned Charlotte de la Tremouille who agreed to transfer the buildings to Sir Humphrey Chetham's feoffees (trustees). Sir Humphrey had been discussing the possibility of acquiring the former college in order to found a new school for several years before he died in 1653 and within a year of his death, the ownership of the buildings was transferred and forty boys were lodged at 'Chetham's Hospital and School'. This was the school and purpose intended by Lord Strange, mentioned in the university petition. Thus was the chance for the University of Manchester to be founded on the banks of an eel and trout river lost for ever.

The 1422 building, whose outline has remained virtually unchanged in centuries, would have made a worthy first College of the University and as a Grade I listed building today, would rank alongside most colleges at Oxford and Cambridge for antiquity and architectural beauty. Henry Taylor [24], in his 1884 reconstruction of how the College appeared when it was built in 1515, (see Figure 1.2), asserted that the functionality of the halls and rooms was similar to contemporary Oxford and Cambridge Colleges.

Figure 1.2: *Henry Taylor's 1844 reconstruction of 'The College', before the Grammar School was built. As Chetham's Hospital and Library, it looks essentially the same today (see Figure 1.3 on the following page).*

The building was perched on red sandstone on the South Bank of the Irk, then a small, brisk, eel and trout river, before becoming a foul channel of domestic and industrial effluent as the city centre became industrialised and overpopulated in the 18th and 19th centuries. The baronial village of Manchester was never given the chance to become a pastoral University market town – a Northern Oxford or Cambridge. The Grammar School, built between 1515 and 1518, would be in the bottom left corner of the picture, nudging the College entrance. I walked through that main entrance in 2010 and took the photograph of the facade across the courtyard (see Figure 1.3 on the next page), looking almost exactly the same as it did nearly 600 years previously, apart from a few drainpipes.

Figure 1.3: *Chetham's Hospital and Library courtyard in 2010. Photo: Robin Marshall.*

Alas, Oxford and Cambridge, King and Church had no wish for any such disruptive scheme to add further disturbance to troubled times and the cultural history of England became thereby the loser, a classic victim of narrow minds with a single self sustaining focus.

This short history of the possible first College within a possible university, serves to reinforce the notion that 1640/1 was not a good time to approach Parliament for such a new university. And even if granted, it is unlikely that it would have survived more than a decade, being under the suffrance and control of the church and government at a time of unstable governance in the nation.

Ferdinando had replied two days after receiving his brother's letter:

'To my very loving brother, Mr Henry Fairfax,
at Ashton-under-Line.

Good Brother,
I have received your letter, and in it a petition for an university to be erected at Manchester, which cannot be done but by a bill in Parliament. The charge will be great – about one hundred marks; and the effecting what is desired will be very uncertain. Those well affected to the now universities (which include, indeed, every member of our House,) will be in danger to oppose this. I should

16

be most glad to have such a bill pass, as beneficial not only to that, but all the northern counties. I shall advise with the knights and burgesses of that county, and go the way they shall think fittest; but I much fear a happy issue of it, especially now that the House has made an order to entertain no new matter till some of those great and many businesses we have grasped be ended, the chief whereof are my Lord Lieutenant's trial, this day only entered into, which is like to hold one week ; the next will be my Lord of Canterbury's trial, and with that.

     ... etc etc ... [Ferdinando then dealt with other business of State and family.]

     Thus, with my best wishes to yourself and my sister, I rest, Your very affectionate brother,

     Fer. Fairfax.'

The Lord Lieutenant referred to was Thomas Wentworth, Earl of Strafford and Lord Lieutenant of Ireland, whose trial began on the 22nd of March 1641 (modern date counting). This date is within days of the exchange of letters between the two brothers on the question of a university for Manchester. It is easy to decide, nearly 400 years later, which matter Ferdinando would have decided was more pressing.

As far as the matter of charge being 'great', the unit of currency in England at that time was based on a silver penny of which there were 160 to the mark and 240 to the pound, the mark often being used for fines and legal charges. Indeed, a cost of one hundred marks was a mere percentile fraction of some of the fines handed down by King Charles' Star Chamber in those days. The 'crime' of criticising Strafford could lead to a penalty of not only many thousands of marks but the branding of both cheeks. Star Chamber fines were, of course, designed to ruin the political opponent, but even so, a bill costing little more than thirteen shillings per head was surely within reach of a hundred nobles, gentry, clergy, freeholders and other inhabitants from those who had signed the petition and who would surely have put their pockets and quills alongside their mouths. This was the difference between the 17th and the 19th century concerning Manchester's aspirations for a university. When it finally happened, it was local money and energy that made it happen.

There was further correspondence after which Ferdinando wrote again to his brother on the 20th of April 1641 (the year is now the same in the Old Style and New Style). He had talked to various gentlemen from Lancashire and Chester, albeit Oxford and Cambridge educated, who were pessimistic about a university for Manchester. He repeated his views on the cost of the Bill, too much to be risked on an uncertain outcome and used local (temporary) matters as a deflection:

> 'my Lord Strafford still keeping us in play . . . . but if there be an open, I shall let you know.'

Historians are unanimous that the country was gripped by a fever in the run up to Strafford's trial. The fever increased during the trial and reached a pinnacle when he was convicted and condemned to death. The Men of Manchester must have been aware of this and their timing appears questionable, nearly four centuries on.

Ferdinando's reply was a classic 'Don't call us; we'll call you' brush off and indeed, the removal of Wentworth's head 22 days after Ferdinando's letter, which simultaneously removed one of the obstacles to the university, did not as far as the Fairfax Correspondence shows, lead to a follow up letter telling his brother vicar, Henry that the chances had improved. Indeed, as fast as one problem was solved, another arose. The peace with Scotland was broken, the Scottish army marched South to seize Newcastle and was even moving on towards Teesside.

That was not the end of Northern attempts to create a new university, and the next quite remarkable step, which happened within a year, included, although undated in the Fairfax correspondence, a virtually identical 'Humble Petition of the Nobility, Gentry, Clergy, Freeholders, and other Inhabitants of York'. Much of the wording of the plea was a clone of that from Manchester, in a few parts more clearly expressed where Manchester had been verbally clumsy. It used the more vague term 'Popery' instead of the substantive 'Papists'. It added an extra reason, in that England might receive some honour by a third university since Scotland (a separate kingdom, coincidentally having the same King at the time) had 'long gloried in the happiness of honour by having the four universities at Edinburgh, Glasgow, St. Andrew's and Aberdeen'. It is possibly no coincidence that Ferdinando represented Yorkshire in parliament and the Yorkist petition used the Mancunian one as a template, especially since Ferdinando had the Manchester petition in his hands.

The rival bids from Manchester and York were sent to the House of Lords in 1641 and then York eventually followed up with a repeat to the Lords and Commons in 1648. Nothing came of either initiative, most likely because England was otherwise occupied in civil war during the intervening seven years. As soon as the conflict was over, Durham entered the field (see for example, J T Fowler [19]).

In April 1649, with the King recently executed, Parliament dissolved all the Collegiate Chapters in England and one consequence thereof was that in Durham, the Bishop and all his subordinates were turned out of house, home, castle and cathedral, leaving an empty College. The Church College in Manchester had already been eradicated and even used as a gunpowder store and prison during the war. The Durham republicans immediately petitioned Oliver Cromwell to permit a new college there and after some written exchanges, a deputation rode over to meet Cromwell, who was on a visit to Edinburgh, in order to plead their case, taking the opportunity to remind him of how Durham and Northumberland had dealt with the Scotch[8] army.

By 1656/7 a college in Durham was founded in law, to which a full transcript of Oliver Cromwell's speech setting up the college, by the writ of Privy Seal can be read in Thomas Burton's diary [25]. Among the staff of mainly religious men listed in this speech, was the eclectic Ezekial (variously Ezerel, Ezereel or Israel) Tonge, a priest, papist plot propagator, botanist and alchemist, thus incorporating a kind of science into the curriculum. Chemistry and alchemy were close relatives at the time. Physics did not exist as a defined subject, but it can be noted that the teacher-alchemist Tonge presaged Ernest Rutherford in Manchester who, after converting nitrogen into oxygen in 1919, designated himself as the first *successful* alchemist. As well as being an alchemist, Ezekial Tonge collaborated with Titus Oates in propagating stories of Jesuit conspiracies and Popist plots. Regarded by all who knew him as somewhere between eccentric and insane, his prose was described by historian J P Kenyon as 'turgid and incoherent'. Yet his 1670 paper to the Royal Society on the motion of sap in sycamore trees is perfectly lucid and readable.

Burton [25] also mentions a proposal for a third English University in London, although this was no more than a letter in 1648 from economist and founder member of the Royal Society, Sir William Petty, to the so-called *'Great Intelligencer of Europe'*, Samuel Hartlib. Petty argued the advantages of learning and that:

> 'Many are now holding the plough, which might have been made fit to steer the state.'

The new college in Durham was an unequivocal, immediate and abysmal failure. Yet notwithstanding its fatal inability to attract students, a

---

[8]All written material until well into the 20th century used the adjective 'Scotch' to describe anything that is now held to be Scottish.

petition was soon sent to Cromwell, seeking full university status and the right to award degrees. Before he could make any decision thereon, Oliver Cromwell died on the 20th of June 1658 (OS). The business was taken up by his son Richard Cromwell and despite the absurdity of unviable student numbers, the patent for the university was drawn up and prepared for Seal. This immediately induced the monopolistic wrath and panic of Oxford and Cambridge who counter-petitioned Richard and even sent personal deputations [25]

> '... to give reasons against a third University, and especially against conferring degrees there, which was much endeavoured by some. Whereupon a stop was put to it.'

Once again, the true reasons for not converting this college into a university did not need to be stated by the Establishment. Reason was not needed, as President de Gaulle noticed in 1967 when for the second time he vetoed British admission to the 'Common Market': if you are in charge, you can just say 'No!'. Thus, Oxford and Cambridge did not object to a Manchester University as such, nor to one at Durham, York, Tamworth or any hypothetical place; they were simply implacably opposed to a new University anywhere. Further national events overtook the decision and Fowler goes on to destroy any credibility for his impartiality as a historian with the remark [19]:

> 'The blessed Restoration, as Carlyle calls it in irony, and as some of us call it in deep thankfulness, put an end to the Durham College.'

This may seem oddly emotive to us now, written 200 years ago about an event that had happened a century and a half before the words were written, but it is no less oddly emotive than York University naming one of their constituents 'Wentworth College', three centuries after the event, honouring a man whom even the meek and impartial historian, John Richard Green [26] described as 'the spirit of tyranny'. As in the US, the civil war in England never ended.

At the restoration, Durham College was abolished and all its republican inhabitants were expelled as unceremoniously as the royalists had been in 1648 to make place for them. Durham College did not become a University and the English university map was rolled up for two centuries. It was a slow realisation, two hundred years later, that new degree awarding institutions had to come from within themselves and not from the Establishment. Ferdinando Fairfax had put his finger on the situation

20

in his first reply to his brother. The 'now universities (which include, indeed, every member of our House,) will be in danger to oppose this'. The 'now' universities of Oxford and Cambridge were an executive arm of a government comprising a tight trinity of King, Parliament and Church. Parliament was exclusively Oxford and Cambridge and Parliament meant the perceived power of those within. Rulers and governments rarely, (South Africa under world pressure perhaps excepted), vote themselves out of power. It is no coincidence that the Representation of the People (Reform) Act of 1832 and the founding of new universities happened within decades of each other, when some of the historic power was torn out of the hands of the privileged.

Thus, a new university in Manchester, or anywhere else in the country for that matter, was a non starter in the mid 17th century. It is nevertheless interesting to speculate on how the university would have fitted into the town which had a population, including surrounding villages of about 5,000. The collegiate Christ Church of the time eventually became the city's cathedral and after 20 years of repair following its Christmas 1940 bombing, looks the same today as it did when it was built centuries ago. The adjacent market place formed the hub of the town, with the existing and 'ancient' college buildings within a few minute's walk. As a college they would have held on to the name 'Christ's College', thus thwarting the carrying away of the name by the body of Clergy who scattered themselves in buildings along Dene's Gate (Deansgate). As a theology college, which all universities were at that time, it would have attracted fewer than 100 students, and it would have been an integral part of both the church and the baronial infrastructure and fabric.

The records of the Manchester Court Leet [9], which was under the ultimate jurisdiction of the Lord of the Manor, cover the period from 1552 (Edward VI) to 1846 (Victoria) when the newly incorporated city purchased the baronial rights from Sir Oswald Mosely. The records offer a glimpse of town life in those days into which a new university would have fitted. Most of the court's time was taken up with enforcing primitive sanitation. Households were responsible for disposing of their own sewage which could be tipped into the formerly clean eel and trout rivers only during night hours. Many failed to do even that, and appeared before the court, to be fined ten groats. Enforcing true weights and measures in the market also featured regularly. Refusing to let the baron's ale tasters taste the house's ale attracted a substantial fine. All this was denied potential university students of Manchester for over 200 years.

Students, like the rest of the population, would have been subject to strict discipline, especially after dark. Those who walked by night and slept by day were automatically regarded as miscreants and reprobates. Playing football was banned because the result was usually the breaking of gentlemen's windows. Plate glass windows were so precious that it was the norm to take them with you when you moved house. Another medieval game, giddy-gordy, held by some to be a precursor of baseball and hockey, was also strictly controlled because the equivalent of the ball or puck was a short cylinder of wood, sharpened to a point at each end like a stubby fat pencil. One pointed end was struck sharply downwards with another stick (club or bat) whereupon the puck flew upwards. Whilst in the air, the aim was to strike the moving puck with the bat, to send it hopefully with intent in a chosen direction, but more likely randomly, into a spectator's eye. This game was also banned by the baronial court, to ensure a peaceful life for the inhabitants.

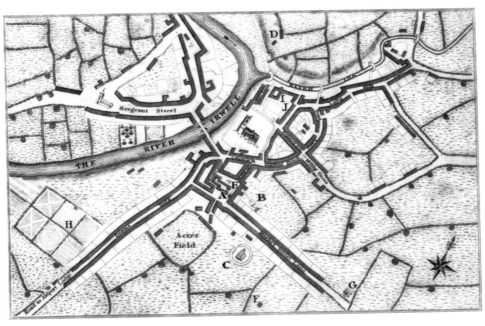

Figure 1.4: *A map of Manchester and Salford around 1650. A: Sessions House. B: Cock Pit. C: Radcliffe Hall. D: Mr Knowles' House. E: Meal House. F: Fountain/Conduit. G: Mr Lever's House. H: New Gardens. I: College. J: Grammar School. Although the College buildings shown here exist almost unchanged in plan view today, the trout stream, the Irk, is now overbuilt by Victoria Station.*

The equivalent of the Old Trafford football ground for village males was a mere furlong from the church and was hence benignly tolerated or even supported by baron and church alike; on leisure days, the crowds flocked to the Cock Pit, to watch, cheer and wager. Thirty two contestants would pay three pounds each to take part. The winning cock would secure for its owner a prime ox, supplied by a dairy farmer in the Yorkshire Dales. Putatively potable water for the town came from a spring which fed the 'Conduit Head' through an elm pipe. The spring and conduit were outside the current town in the area of Manchester where Fountain Street and Spring Gardens are now to be found. Piccadilly was then an open field. The Court Leet records regular fines and court orders imposed on those who misbehaved at the 'Conduit Head'.

Figure 1.4 on the facing page shows Manchester around 1650 including the locations of several places that have been mentioned here. This is one of several versions of the so-called 1650 map, appearing in Palmer's History of the Siege of Manchester [23] and has the fewest errors. When laid over a properly surveyed map of the 18th century, approximations to some of the streets and of the junction of the Irk and Irwell can be noticed. The map was originally published inverted, viewed looking South; the present campus of the University being a considerable distance off the top of this image. For reference, Acres Field near the corner of Market Street and Deansgate is now St. Ann's Square.

If the experiences of the colleges which sprang up in the region just over a hundred years after the failed Manchester petition are anything to go by, there would eventually have been problems with boisterous rowdyism as the students celebrated every English defeat in the American war of Independence. All of the first four King Georges were never popular among students.

With its aspirations for a university dashed, the village of Manchester slipped in and out of a century and a half of slumber, virtually untouched by the creation of the United Kingdom in 1707. Then suddenly, Manchester became the hearth of the Industrial Revolution and a new generation of ambitious and technologically minded businessmen transformed the baronial village into a heaving, filthy city. Eels and trout disappeared from the Irk and Irwell, to be replaced by stinking industrial effluent. The chance of a Northern pastoral University town, to match Oxford or Cambridge, had gone.

A feature of the 100 years that followed the civil war was the birth and proliferation of so-called academies, which were designed to produce priests for the growing number of dissenting and non-conformist churches. A feature of the academies was the diversity of their curriculum which included the sciences. As Manchester grew into an industrial city, the notion of a college to support the needs of the city arose once again and several emerged in the space of only a few years, some disappearing as quickly as they appeared. Two of them, both secular, survived to become the eventual University of Manchester, whereas all the colleges founded on theology eventually vanished. They were all intertwined to some extent, often with common trustees and we shall now look at a brief history of each of them in turn. But before these Manchester born colleges and academies were created well into the 18th century, Manchester was touched briefly by an academy of quality that had arisen in the Yorkshire Dales, near Settle.

1669–1712: The Rathmell Academy and Manchester

The indirect trigger for the appearance of the first academy for young men in Manchester was the restoration of King Charles II in 1664, and the various actions and Acts that resulted in the ejection of approximately 2,000 ministers, (see Edmund Calamy [27]) from the Church of England because they refused to 'conform'. Cut off from the established church and its colleges, many of those ejected or silenced ministers with a mission to teach, set up their own unofficial academies, often in their own houses, where the next generation of non-conformist priests could be spawned.

RIC° FRANKLAND.   1630 - 1698

Figure 1.5:

Richard Frankland, the excellent teacher of future teachers of physics in Manchester.

This portrait by Gustavus Ellinthorpe Sintzenich was made between about 1841 and 1892 from an earlier painting.

This image is declared by Wikimedia Commons to be in the public domain.

One such dissenter was the highly respected Richard Frankland and he established a vigorous academy in the hamlet of Rathmell in the Craven district of Yorkshire in March 1669. A portrait of Frankland, done about

150 years after his death, is shown in Figure 1.5 on the preceding page. The word 'college' then had a strict ecclesiastical meaning and I use the word 'academy' here for anything not connected to the established church. This academy, situated between Settle and Hellifield, taught science as a minor adjoint to dissenting theology. Its founder Richard Frankland had been involved in the short-lived attempt to found a university at Durham. Of his association with Durham, Calamy was moved to remark [27]:

> 'Mr Frankland was pitch'd upon as a very fit man to be a tutour there.'

The word 'tutor' in those days was used for a person of equivalent professorial standard but who did not hold the chartered title of professor.

If Rathmell seems a remote place to establish a college, both Giggleswick School near Settle and Ermysted's Grammar School in nearby Skipton, had been educating boys to (theological) Oxford and Cambridge entrance standard since 1512 and 1493 respectively. Frankland himself went to Giggleswick School before gaining a scholarship to Cambridge, and his new college offered a local, cheap alternative for non-conformists. I went to Ermysted's.

From 1669, sheltered from the northerly winds by Ingleborough, Penyghent and Whernside, the college's students, including James Clegg [28], went through a course of 'Logick, metaphysicks, somatology, pneumatology, natural philosophy, divinity and chronology'. Most of these terms had meanings that differ from what a physicist might expect today, e.g. pneumatology has nothing to do with the kinetic theory of gases, but more to do with the holy spirit. Metaphysicks was concerned with the philosophy of existence, objects and their properties, space, time, cause and effect, and even probability. Natural philosophy was a subset of metaphysicks but began to emerge as a subject of its own as practical experimentation developed. In 1669, Newton was 26 and his *Philosophiæ Naturalis Principia Mathematica* was 18 years into the future. Students who wished to obtain a degree could do so in Scotland after attending a single session at Rathmell, such was the quality of its teaching and learning. Clegg, more of whom later, obtained Scottish degrees, including medicine and eventually combined the duties of non-conformist minister and general practitioner in Chapel-en-le-Frith.

Frankland endured relentless harassment and persecution for operating his 'illegal' non-conformist colleges and he survived by a combination of determination, diplomacy, itinerancy and by locating his colleges outside the five mile exclusion zone, which surrounded Anglican churches. Not long after he had established his college at Rathmell in 1688, he took it to Natland near Kendal in 1674, from whence it went to Kirkby Malham in 1683. There followed a sequence of relocations back to Kendal, then to Windermere, followed by Attercliffe in 1686 before finally returning to Rathmell again in 1689 where it stayed until Frankland died in 1698. His death in 1698 signalled the end for the thriving Rathmell Academy. One of his former students, Salford born (1666) Rev John Chorlton, having declined to take over in Rathmell, took a dozen of the students with him and set up an establishment for 'university learning' in a great house in Manchester instead.

There are no clear records of where this 'great house in Manchester' actually was. A likely location (but still a guess) would be the 'Dissenters' Meeting House', which was the name of what became to be called Cross Street Chapel, built in Pool Fold Lane, running along the hedgerow on the left hand side of Acres Field in Figure 1.4 on page 22. This non-conformist chapel was built in 1694 for the former pulpit holder of the main church, Henry Newcombe, who was himself ejected. The building could seat 1,515 persons and in 1713, Calamy [27] described it as 'A large stately Chapel on the South-Side of the town, called Ackers'. It was initially called the 'Dissenters' Meeting House' and fits the bill to be a 'great house'. The site for the meeting house was on the edge of Plungeon's Field, which contained the village's ducking pond and stool. It may or may not have housed this small academy although it was definitely the subsequent home of two later academies.

Another possible site near the chapel would have been the moated Radcliffe Hall on the other side of the ducking pool, shown as 'C' in Figure 1.4 on page 22. The Radcliffe family were staunch parliamentarians but had been wiped out, apart from the youngest daughter, by the plague of 1645. The house had been used as a prison during the civil war and so was easily big enough for classes of ten students. However, this is all speculation without further evidence and all that can be said is that physics was being taught in this area of the village of Manchester between 1699 and 1712.

The 1741 map of Manchester, produced by Casson and Berry, is the first one available after the 1650 map (which was shown in Figure 1.4) and the section displayed in Figure 1.6 on the facing page shows the relatively small amount of development south of Market Street in the 90 years between the two maps. Acres Field has become St Ann's Square and the new St Ann's Church (36 on the map) and the big chapel (35 on the map) are within a few paces of each other along Queen Street.

26

John Chorlton died in 1705 and his colleague, John Coningham kept the academy going in Manchester until 1712 when he was persecuted and prosecuted by the established Church and State for running it. Being of a timid nature, he immediately relocated himself to London. This was the end of Manchester's first academy, but not the end of what Frankland had started.

One of Chorlton's students in Manchester, Thomas Dixon (1700–05), gained an M.A. in Edinburgh (1708) *honoris causa*, answered a call to lead a chapel in Whitehaven and set up his own academy there in 1710. Lecture notebooks written by one of Dixon's students Henry Winder have survived [29] and are held by Harris Manchester College, Oxford, where they are safe. They contain two lectures on mathematics and astronomy – in Latin, taught and learned in Manchester and taken to Whitehaven.

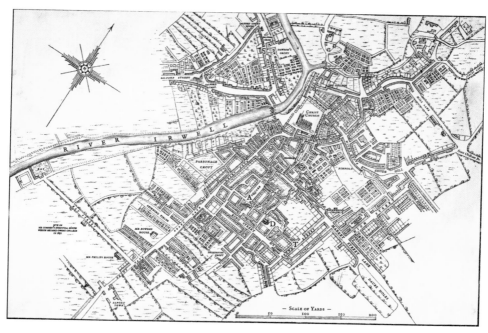

Figure 1.6: *A map of Manchester in 1750, probably traced from Casson and Berry's 1741 survey. A: St Ann's Square, formerly Acres Field. D: Dissenters' Meeting House in Pool Fold Lane, later renamed Cross Street Chapel in Cross Street. T: The Theatre.*

The sequence of irrefutable facts related above, concerning the location and leadership of the various academies that sprang from Rathmell, show beyond any doubt that the claim by Harris Manchester College to be directly descended from Rathmell Academy is completely without foundation. Even the erudite Oxfordian Irene Parker dismissed the claim in her 1914 book [30]. The pretending claims to direct descent made by Harris Manchester College in the proceedings of its opening ceremony in 1893 [31] are simply wrong and being repeated by Davis [32] in 1921, did not make them right. Rathmell College, created and guided by the honest Frankland, who had intelligence, vision, energy and acumen, and who demanded respect from his religious enemies, ended when he died. The Manchester Academy was not a sequitur of Rathmell; seven eighths of Rathmell's students dispersed elsewhere. The Whitehaven Academy started two years before the demise of Chorlton's Manchester Academy. Harris Manchester College was eventually spawned by the spontaneous creation of the Warrington Academy, which landed in Manchester in 1786, on its way to Oxford, thus inserting the word Manchester into its name.

Thus it came to pass that the teaching of natural philosophy and physics gained its first toe-hold in Manchester in 1699 and clung on for a further baker's dozen of years. After this stuttering start, nothing much happened for another 70 years as Manchester went to sleep again, natural philosophically speaking.

1720–1799: Wandering Natural Philosophers

After the briefest flicker of life for physics in Manchester – at university level, no less – there was no more formal college or academy teaching in the village until the academy of the new Manchester Literary and Philosophical Society was opened in 1783. This fallow period in Manchester coincided with the spectacular advances made in the subject of electricity and magnetism. Physics has had many golden ages, and this was one of them. A comprehensive account of electricity during the 17th and 18th century has been written by Heilbron [33] and is well worth a read.

On the 2nd of April 1747, the former Rathmell student James Clegg, by now occupying a non-conformist pulpit in Chapel-en-le-Frith, some 20 miles from Manchester, made a laconic entry in the diary [28] he kept throughout his life:

'(At Manchester.) We saw the Electrical Experiments.'

Clegg had a tendency to be accident prone; he regularly fell out of the saddle of his stumbling horse, was frequently poked almost mortally in the eye by trees as they were being felled, he was violently thrown by the horn of a madding cow and he even set fire to his own bed with candles whilst drowsing in it. About these events he waxed verbally in his diary, praising the Lord for regularly saving him. But he saw nothing exciting in electricity. Yet his response to what was clearly a steam engine, 17 years before he saw a demonstration of electricity, was inspirational:

> '28/9/1730 came to Winster about noon. Saw 3 curious Engines at work there, which by ye force of fire heating water to vapour, a prodigious weight of water was raised from a very great depth and a vast quantity of lead oar laid dry. The hott vapour ascends from an iron pan close covered, through a brass cylinder fixed to the top, and by its expanding force raises one end of the Engine, which is brought down again by the sudden introduction of a dash of cold water, into ye same cylinder which condenseth the vapour. Thus the hott vapour and cold water act by turns and give ye clearest demonstration of ye mighty elastic force of air.'

In his 1811 report, [34] for the Board of Agriculture, John Farey, assuming that all of his readership knew about galena, Crown property no less, wrote:

> 'A few years after their introduction, 10 Steam-Engines were erected in the immediate neighbourhood of Winster.'

With his excellent mathematical training and some thought, Clegg could have calculated how many pounds of best Derbyshire coal needed to be burnt in order to lift a pound of lead ore through one foot and thereby discover a century before Joule, that the number was essentially a universal constant (for the same sort of engine and same sort of coal). Instead, he focussed on the elastic force of air, which was nothing more than the mechanical coupling agent between the energy and the water to be lifted.

In the same year, the energetic and entrepreneurial John Wesley, a mutually respected associate of Clegg's, who also kept a diary, noted his own attendance at a lecture on electricity [35]:

> 'Fri 16 Oct 1747
> I went with two or three friends, to see what are called electrical experiments. How must these also confound those poor half-thinkers, who will believe nothing but what they can comprehend? Who can comprehend, how fire lives in water, and passes through it more freely than through air? How flame issues out of my finger, real flame, such as sets fire to spirits of wine? How these and many more strange phenomena arise from the turning round a glass globe? It is all a mystery.'

29

Unlike Clegg, Wesley was not satisfied with not understanding and set about educating himself. He eventually used a precursor of electric shock therapy in the several clinics that he established around the country.

1747 saw a surge in public lectures on science, caused almost entirely by the invention of the capacitor. In 1745, Lutheran cleric Ewald Georg von Kleist, whilst Dean of the cathedral in Prussian city of Kammin (now Kamień Pomorski in Poland) had invented the Leyden jar, which for the first time, permitted the storage of significant amounts of static electricity. Von Kleist even knew exactly what he was doing. Glass did not conduct electricity so he might be able to fill a glass jar with it. He was right.

Professor Pieter van Musschenbroek of Leiden invented his similar jar in 1746 and tested it on himself. He described it in a letter to his friend, French scientist René Antoine Ferchault de Réamur and advised:

'I would not take a second shock for the Kingdom of France.'

Electricity dominated science in the 18th century which had begun well when Newton's laboratory assistant Francis Hauksbee had taken his master's advice, changed the sulphur sphere in his friction machine for one of glass and produced a highly efficient and portable electrical generator. By connecting the machine across an evacuated vessel containing a dash of mercury, he could make the vessel glow brightly in the dark. Without knowing it, he had invented both an electron accelerator and a mercury discharge tube at a stroke. These new friction machines combined with a portable storage device, the Leyden jar, enabled the best lecturers and demonstrators to go on tour. Lectures were held in theatres, pubs, coffee houses and drawing rooms of large houses. During the winter, outside the main touring season, many lecturers held courses in their own homes.

Amongst the middle and upper classes, the possession of an electrical machine for after dinner entertainment was irresistible. There were regular articles and books on the subject [36] such as John Neale's *Directions for Gentlemen who have Electrical Machines* and *Seventeen electrical experiments for a gentleman to perform with plants and animals*. Given the tendency to electrically charge and discharge women rather than animals for entertainment, the latter article was perhaps coyly titled.

These lectures were extremely popular throughout Europe in the middle of the 18th century although their popularity had faded by the last decade of the century, possibly because like a movie, there is a limit to the number of times an average person can watch the same performance of an average movie. The Revolution in France was also a distraction. The frontispiece to Abbé Jean-Antoine Nollet's book on electrical experiments [37] is reproduced in Figure 1.7. It shows a demonstration of electricity at Versailles. An insulated and suspended woman, having been electrically charged by a rod, is about to have a spark induced from the end of her nose. Sometimes, a spark from the end of a finger was used to ignite a bowl of alcohol.

Figure 1.7:

A public demonstration of electricity by Jean-Antoine Nollet at Versailles around 1750.

With only two universities in England, neither of which felt it within their compass to educate the general public, a large cohort of entrepreneurial lecturers became itinerant because there was money to be made. At their peak, there were about 50 of them touring the country, with Bath being a popular venue. Without getting into arguments about historic inflation, a decent set of electricity demonstration equipment incurred a one off capital cost of £300 in 1747, which was the worth of fifty horses (fit mares) according to Clegg's diary entries (he got through a lot of horses since although he survived tumbles and falls, his horses didn't). This was about the same as the annual salary of a natural philosophy professor. But

31

the best of the touring lecturers could earn over £1000 in one year, properly equipped. I know of no physics professor today who could afford to buy over 150 decent fit horses a year, every year.

A typical set of electrical equipment can be seen in the printed lectures on Natural and Experimental Philosophy by George Adams [38], reproduced in Figure 1.8. The set includes the obligatory friction machine and Leyden jar. The sheer scale of the equipment required was another factor in the decline of the touring lecturers: as the subject developed the amount of the necessary equipment would no longer fit onto a horse and cart capable of traveling the length and breadth of the country. The largest set of equipment on record weighed one and a half tons, requiring more than one horse. Physics had become a circus.

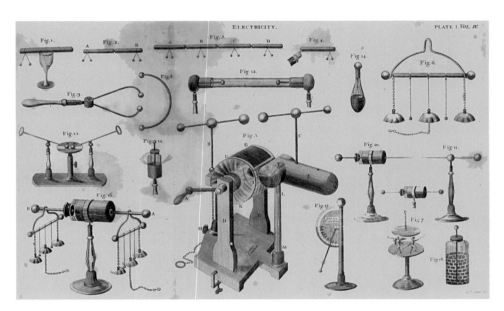

Figure 1.8: *A typical set of mid 18th century electrical equipment for use in lectures.*

An account of the lecturers and their lifestyle can be found in Musson and Robinson [39]. Adam Walker was one such lecturer, including Manchester among his venues during the early part of his touring lecture career, even making it his base for a few years. Walker was only 16 and had not yet started touring when Clegg went to see the electrical experiments in 1747 and Clegg did not record the name of the lecturer, despite meticulously noting the names of everyone he encountered on his daily rounds in Chapel-en-le-Frith, especially the many women who gave him free teas.

After a few years touring the country, Walker made Manchester his home base in 1763 and he lectured there for three years. According to Musson and Robinson [39], he used 'The Theatre' as the venue. The first theatre on record in Manchester opened in 1755 and this is Walker's likely venue for his experimental and natural philosophy lectures. The theatre was built on the corner of Brown Street and Marsden Street, its eventual location denoted by 'T' in Figure 1.6 on page 27. It was close to the Dissenters' Chapel and St Ann's Church in what was now the fashionable part of Manchester, South of Market Street. Eventually, one hundred years after Clegg saw the electricity experiments, Joule would present his pivotal discoveries in the same area – the reading room of St Ann's Church.

Figure 1.9: *A Walker lecture on astronomy at the English Opera House.*

With a whole theatre at his disposal, Walker put on a splendid show. When he moved to London, he used the Theatre Royal, charging two or three shillings a show, half price for ladies. The engraving shown in Figure 1.9 is usually designated as Adam Walker lecturing on astronomy at the (New) English (Covent Garden) Opera House in 1819. But the animated, dark haired figure is hardly the 88 year old Adam Walker, but almost certainly his son, David Franklin Walker, who followed in his father's footsteps. It was patriarch Adam Walker who invented the eidouranion, a back-illuminated transparent orrery, of which a 27 foot diameter version can be seen in Figure 1.9 on the previous page. The theatrical lecture was accompanied by music from a modified harpsichord that Adam Walker also invented.

Walker wrote up his lectures into a book [40], which he had printed and published under his own name. The first of 10 editions describes him as a 'Teacher of the Belles Lettres in Manchester'. The arrangements for the lectures were given at the end:

> 'Each of the twelve Lectures requires about two hours attendance, and begin in the country so soon as forty have subscribed one guinea each, to be paid at the first lecture, when the days and hours of attendance for the future lectures will be fixed agreeable to the majority of the subscribers. In Town, the Lectures are every Monday, Wednesday, and Friday, at twelve, at Mr Walker's house, George-street, Hanover Square. Noblemen, or others, may have this course read at their own houses, (if at a proper distance from the apparatus) on easy terms.'

Adam Walker became famous and wealthy and was the subject of the caricature shown in Figure 1.10 on the facing page drawn by James Gillray in 1796. Gillray, perhaps remembered most for his caricature of Pitt and Napoleon carving the plum pudding globe, could be vituperatively cruel and crude in his cartoons, especially with royalty (George III and Charlotte) and politicians (Pitt). But his sympathetic portrayal of Walker shows a man at the peak of his skills and a scientist in love with his science. The miniature inserts at the top of the picture show Walker's mentor Joseph Priestley on the right, facing Walker himself on the left.

In the middle of 17th century, at the time of the civil war, life in Manchester was centred North of Market (Lane) Street with the Established Church at the core of the village. 100 years later, the surge of independent faith and an interest in science had taken root in the region around Cross Street, which at the time was still called Pool Fold Lane near Market Street, Cross Street near the Dissenters' Chapel and Red Cross Street near where what is now Albert Square. Yet despite being briefly touched by the physics of electricity, Manchester itself contributed nothing to the advancement of the actual science. Even so, the spectacle of

Figure 1.10:

Adam Walker lecturing – a cartoon by James Gillray.

electricity fired local interest and enthusiasm and it was not long before the interest in science, added to that of the Arts, generated sufficient motivation to form a Society and a College in the newly fashionable hub around St. Ann's Square and Cross Street.

1781: The Literary and Philosophical Society

The founding of the Manchester Literary and Philosophical Society, or the 'Lit. and Phil.' as it has always been known, on the 28th of February 1781 was a landmark in the evolution of science in what was still a baronial manor village. It provided a forum for discussion on the science and arts and perhaps more importantly, it allowed practising scientists who did not have access to a friendly Fellow of the Royal Society who would present their papers to the Manchester Society in order to publish their work. The Society immediately began publishing its memoirs containing presented papers and among them are several of an epoch making nature, such as those by Dalton, Joule and Rutherford.

On its foundation, the Society was closely associated with the Dissenters' Chapel and held its meetings there. The Society met every two weeks and provided a structure for the presentation of scientific papers and their critical assessment, both by peers and by those members of the local community who were interested in science. It carries on this contact with the scientific lay community on a regular basis to this day. In its early days, John Dalton was a regular contributor after being invited to join the Society in 1794, becoming its secretary in 1800 and then president in 1817. He launched the society on a course of excellence. James Joule became president in 1860. The close working connection between Joule and William Thomson, (later Lord Kelvin), ensured that Thomson was a regular attender and contributor to the meetings.

In 1799, the Society secured its own premises at 36 George Street (see the montage in Figure 1.11) and stayed there until the building was destroyed in 1940.

Figure 1.11:

36 George Street, Manchester. Vice-President, Peter Ewart and President, John Dalton discuss on the pavement.

Colour montage © 2017 Robin Marshall.

One might have expected that when Owens College opened up in 1851, their first three physics professors, Archibald Sandeman, Robert Bellamy Clifton and William Jack might have availed themselves of the Society's pages, but they are conspicuous by their relative absence. It was the physics professorial duo, Scotsmen Balfour Stewart and Thomas Core who appear regularly from 1870 onwards.

Indeed, physics itself was slow to get a toe-hold in the Society's journal and what papers on physics there were in the early days were often simplistic reviews of well established topics. Apart from John Dalton and James Joule who eventually strode their respective worlds like Collossi, a few others deserve mention and recognition for their contribution to physics in the village and I pick out John Gough, Peter Ewart, James Bottomley and James Nasmyth, all of them regular contributors, for justified attention. Manchester resident Peter Ewart's contribution to physics and its understanding was so striking that I defer a description until the section on Joule and conservation of energy, later on in this chapter. Likewise, James Nasmyth is worthy of special attention.

Scotsman James Bottomley was also a prolific publisher in the memoirs on physics topics. Although not a resident of Manchester, he was the first choice to be Balfour Stewart's professorial partner in Manchester in 1870, when the endowment and creation of the Langworthy Chair enabled the Physical Laboratories to expand enough to justify a second chair. He turned Manchester down and went to Glasgow University instead.

Although John Gough is little known outside his special sphere of influence, he deserves consideration and admiration for sustaining a career as a scientist in many fields despite being blinded by smallpox when he was five. He grew up in Kendal and was a tutor to John Dalton, teaching him Greek and Latin. To this classics knowledge he demonstrated by his 14 papers published the Society's journal that he also skilled in zoology, geology, acoustics and botany. As a child, blind, he had been much perplexed by the sound of repeated multiple hammers as they were wielded by carpenters. The sounds and his reaction to them were accentuated as his viable senses strove to compensate for the loss of sight. Reminded of the percussion of hammers in adulthood, he wrote a paper for the Lit. and Phil.'s journal where he was able to transpose the word 'hammer' into prose in the title of his paper: *The laws of motion of a cylinder, compelled by the repeated strokes of a falling block to penetrate an obstacle, the resistance of which is an invariable force.*

But before that paper, he published one of singular priority in 1805 [41], marking himself as the first polymer physicist to make a serious contribution to the field. His paper, entitled *A description of a property of caoutchouc or Indian rubber; with some reflections on the cause of the elasticity of this substance.* Several years earlier in 1770, Joseph Priestley, in his classic book on the art of perspective drawing[9], as an addendum to the Preface, drew attention to the fact that a substance excellently adapted to the purpose of wiping from paper the marks of a black-lead pencil. It was sold by Mr Nairne, mathematical instrument-maker, opposite the Royal Exchange. Nairne sold a cubical piece, of about half an inch, for three shillings[10] and said it would last several years. Edward Nairne operated his business from 20 Cornhill in London and serendipitously discovered the superior property of caoutchouc over breadcrumbs which had been used till then. Reaching out for a piece of bread to correct a mistake, he accidentally picked up a piece of rubber and immediately started selling what was until then a useless curiosity in Europe.

Gough described three experiments that he carried out on a prepared sample of the resin, as he called it, to establish its singular property. His first experiment established that if the resin be suddenly stretched, its temperature rose and the more it was stretched the more its temperature rose. This was completely contrary to virtually all other solid substances put under tensile stress, whose temperature fell as heat energy was used up to overcome molecular bonds. His second experiment established that the material had a negative coefficient of thermal expansion, a property possessed by few materials. His third experiment, at first sight, would appear to serve no useful purpose to the casual observer. If the thong of caoutchouc, as he now described it, were stretched in warm water, it retained its elasticity as defined by Hooke. But if it were stretched in cold water it would not contract when the stretching force was removed. But if warm water were now added, it immediately contracted. From these observations Gough concluded that the elasticity of the material was not a constitutional quality of the substance, but a consequence of the fact that the resin lost equilibrium with the portion of *caloric* that it possessed at the moment of investigation and its capacity to receive more when needed. Gough was not alone in those days, making scrupulous correct observations, but failing to interpret them correctly due to an implacable faith in phlogiston or the æther.

Gough could not, of course, have possibly anticipated that a full interpretation of the phenomena that he had meticulously recorded required a complete knowledge of science yet to be formulated. The

---

[9]Any aspiring artist believing that geometry forms no part of art, is well advised to read Priestley's book, which is still widely available. Priestley dedicated his technically artistic book to Joshua Reynolds.

[10]Its equivalent, £12 in modern money is not completely outrageous for state-of-the-art materials.

physical properties of rubber and other polymers like it are determined by the fact that it is composed of a highly entangled chain of very long polymers. In that configuration they find a local minimum in total energy. One hundred and sixty years later, Professor Sam Edwards in the Physical Laboratories in Manchester resolved to put the whole behaviour on a sound physical basis, and he was fortunate that he had the tools of thermodynamics, statistical mechanics, entropy, Gibbs free energy and a fertile brain to produce a theory that dealt with Gough's correct observations and many more since. Sam Edwards appears in Chapter 8 of this treatise.

1783–1785: The College of Arts and Sciences

On the 6th of June 1783, the two year old Manchester Literary and Philosophical Society instituted its own 'College of Arts and Sciences' based in the Chapel, hoping to offer advanced courses to those unable to graduate from Oxford and Cambridge on religious grounds. The constitution of the college, the syllabus and lofty aspirations were published in the 2nd volume of the Society's memoirs [42]. Thomas Barnes, minister of the Dissenters' Chapel in Cross Street supplied the motivation for an 'Extension of liberal education in Manchester' especially the *desideratum* which many parents would be happy to see supplied, to solve the problem encountered by those young sons who 'were rather too old to pace round the beaten track of a grammar school; and yet too young to be trusted abroad in the world as his own master'. It was therefore proposed to establish in Manchester, a seminary of liberal science. It was further proposed that several gentlemen shall unite in delivering a course of liberal instruction in Languages, the Belles Lettres, History, Commerce, Law, Ethics, Natural Philosophy, Chemistry and Mathematics. A 'rude and general' outline of the syllabus was given:

> 'Natural philosophy in all its branches, (except Chemistry), including Optics, Pneumatics, Hydrostatics, Astronomy, Electricity, &c. or the discoveries relating to Vision, Air, Water, the Heavenly Bodies, Electric Fire, &c. and attended with experiments on Microscopes, Telescopes, Air-Pumps, Fire-Engines, Orreries, Electric Machines, &c. will form a very large and important part of the proposed plan.'

A 'fire-engine' was the first steam engine to be applied industrially, designed by Thomas Savery in 1698.

The target age group of the young men of Manchester was 14–18, most of whom at the time were already in employment. Thus the concept and execution of this college would not add to the knowledge of physics, but merely propagate existing knowledge to the youth of Manchester. The lectures were delivered in the evening, so as not to interfere with the regular hours of business. Tuesday, Thursday, Friday and Saturday were the days appointed for lectures and the time of lecturing was from six till nine pm with intermissions –'... about half an hour, or an hour. No lecture should usually exceed an hour with no more than two lectures to be given per evening'.

The first report of the college [42] contains the entry that the course of lectures on practical mathematics and the principal branches of natural and experimental philosophy, due to be presented by Mr Henry Clarke was not, in fact delivered. Thereupon, this whole enterprise was short-lived, closing in 1785 due to lack of support. Lack of support was probably the reason why the lectures on Natural Philosophy were not delivered and there is a stark contrast between this college being unable to attract, say, even a dozen young men to attend lectures and the ability of Adam Walker to fill theatres with adult gentlemen and their wives, who watched his physics showmanship in awe. The 3rd volume of the Lit. and Phil.'s *Memoirs* did not appear until 1790 and contained no mention of the new, but already defunct college. Henry Clarke was a professor of experimental philosophy at the Royal Military College in Marlow. His book *A Mathematical and Philosophical Glossary* is listed in Vol 1 of *The Mathematical Repository* [44] as being 'Nearly ready for the press, and will be published in the ensuing year'. Already sufficient data are on the table to have doubts about Henry Clarke's ability to deliver, especially a lecture course in grimy Manchester when he was based on the leafy banks of the Thames in Buckinghamshire where he had to meet daily demands from the military, without having to worry about being sent to fight Napoleon.

A large number of highly skilled mathematicians and natural philosophers were in employment at military colleges at the end of the 18th century. Henry Clark's pending book, like his lectures, was not delivered, although his military college colleague, Peter Barlow, published a superb volume [45] *A new mathematical and philosophical dictionary* in 1814. Reading these books, it is noticeable that by 1814, the long 's' had been removed from printers' fonts.

At a time when would-be colleges and academies in Manchester could hardly fill a dozen seats, it is profoundly instructive to ponder an aquatint of the Chelsea Military Asylum College in 1806 (Figure 1.12) with hundreds of children of soldiers of the regular army studying from nine till five whilst would-be officers were cramming themselves with mathematics and physics at the Royal Military Academy at Woolwich. The Academy regularly published sets of exam questions and answers in the *Mathematical Repository* [44], supplied by professional and amateur mathematicians around the country, mostly using pseudonyms like 'Quinbus Flestrin', 'Sir Titus Triplicate' and 'Theodosius'. The questions on complicated algebra and geometry, answered by these nascent officers would defeat most teenagers today and many questions could not be answered without a firm knowledge of calculus. I would struggle to answer the question from 'Edinburghensis', to calculate the angle at which a horse must tilt when galloping at a speed v in a circle of radius r: just the thing needed to skirt a flank of Bonaparte's army, or at least to provide an understanding why it was risky to force the horse to stay upright if you wished to remain in the saddle. Were horses fitted with speedometers? One must assume that proficient officers were capable of judging the speed of a horse with a precision to match their knowledge of Newtonian mechanics and dynamics. No such pictures exist of classes in 18th century academies in Manchester, because they were not worth painting.

Figure 1.12:

Chelsea Military Asylum College in 1806.

Image taken from the splendid book 'Microcosm of London' [43].

The Society continued to meet in the Chapel for fifteen years after its founding, before seeking its own premises. The records of the Society, published regularly as 'Memoirs' and 'Proceedings' contain a permanent record of the physics research eventually carried out during the early years of Owens College, following its founding in 1851. At the society's meetings, local professors and others were able to make short informal presentations on new ideas or lines of research, some of which were never heard of again. Sometimes the topic reappeared in a fuller form after more work. Ernest Rutherford chose the Lit. and Phil. forum to make the first public presentation of his new theory of the structure of the atom in 1911.

As the newly established Lit. and Phil.'s Arts and Science college faded into oblivion, the small amount of physics teaching in the area, if any, was taken up by a foundling, born out of the closure in 1782 of the Warrington Academy which itself had been founded in 1756. As fast as one college teaching science closed, another opened. The Rev Thomas Barnes, the minister at the Cross Street Chapel had been tireless in helping to set up the Lit. and Phil.'s 'College of Arts and Sciences' and as it declined and died, he directed his energies into creating the successor to the Warrington Academy in Manchester. Barnes was regarded as 'a man of great ability, with a mind of varied culture, and a manner and address that won him numerous admirers' [1]. Be that as it may, the several colleges that he instigated in Manchester had a tendency not to endure.

1786–1803: The Manchester Academy or New College

The Manchester Academy grew out of its direct predecessor, the Warrington Academy, which had been active as a teaching establishment from 1756 to 1782, having been set up, as usual, by those who dissented from the established Church of England. A brief history of the Academy was written by Bright in 1859 [10]. At the Warrington Academy, a young Joseph Priestley, theologian and natural philosopher, had taught modern languages and rhetoric, although not the science he might have preferred. He and fellow tutors are widely held to have set standards such that Warrington Academy became a university in all but name, although this was long before Priestley made any of his scientific discoveries and had attracted the wrath of the Establishment through his implacable views on the corruption of established Christianity. His tenure at Warrington preceded his scientific fame.

Like all other similar establishments in the North West of England, apart from Rathmell, the Warrington Academy had encountered difficulties due to inevitable low student numbers, leading to financial problems and its eventual closure in 1782. Poor student discipline was a regular problem and coupled with endless unresolved debates about the underlying principles of education, this led to a loss of confidence from its financial backers. It was formally dissolved in 1786 and as the Warrington Aca ended, so the Manchester Academy began in the same year, as a direct successor, receiving Warrington's substantial library as well as sharing the residual funds with a new college established at Hackney. As a dissenting college, it taught radical theology as well as modern subjects, including the sciences and modern languages. But the ubiquitous Thomas Barnes, who had been put in place as Principal, made it clear that its main objective was to provide 'a full and systematic course of education for students for the ministry' with a secondary aim to prepare students who wished to enter the medical and legal professions. Any physics being taught was therefore barely even ancillary to the main aims of theology, to which health and law were already secondary.

There are several reasons given why the Academy moved on to York after only 17 years in Manchester. The Unitarian minister Charles Wellbeloved, resident in York, was sought after to become the next Principal but he declined to move to Manchester. Therefore the Academy, it is said, moved to York. There is some credence to this by the fact that when Wellbeloved retired in 1840, the Academy moved back to Manchester. It is also possible that the Academy felt safer in distant York. In 1791, the rising attacks on one of its Warrington Academy alumni, Priestley, culminated in a violent mob burning down two non-conformist chapels in Birmingham, where Priestley was now living. His own house and those of other dissenters were set alight and he lost his laboratory and all family belongings. On the 4th of June 1792, a like-minded mob in the equally riot-prone Manchester had uprooted trees in St Ann's Square and tried to batter down the doors of Cross Street Chapel as well as those of the new dissenting chapel in Mosely Street.

Modern historians tend to concur with contemporary opinion that the riots were condoned, even planned by local magistrates, and were likely to have been encouraged by the King (George III), who made his views on enlightened scientists perfectly clear in a vituperative letter [46] to the Home Secretary, Henry Dundas:

'I cannot but feel better pleased that Priestley is the sufferer for the doctrines he and his party have instilled, and that the people see them in their true light.'

Shortly after, magistrates and government were again heavily implicated in the massacre in St Peter's Field in Manchester on the 16th of August 1819 when the Anglican clerical magistrate, the Rev Charles Ethelston read the Riot Act before ordering the militia to 'control' the peaceful crowd, which they did by killing ten of them on the field and one as he tried to enter his own home. A further 350 were grievously injured, many with limbs and in the case of women, breasts, severed by a drunken, rampaging corp of the Manchester Yeomanry. Thereafter, this regiment was never able to appear in public again without provoking hostility and it was disbanded in 1824. Its place in the historical record of the nation is singularly associated with the killing of its own townsmen and townswomen. Henry Brougham, later Lord Brougham and Vaux, a liberal politician and progressive educationalist stated his views on the root cause of the Peterloo atrocity in a letter to Earl Grey, dated the 31st of August 1819:

> 'The magistrates there (in Manchester) and all over Lancashire I have long known for the worst in England, the most bigoted, violent and active.'

Opinions on the massacre range across the political spectrum and Brougham's can be calibrated by his opinion on slavery, which despite being unlawful in England in 1819, was still more than tolerated by those with a commercial interest in its continued existence:

> 'The slave ... is as fit for his freedom as any English peasant, ay, or any Lord whom I now address. I demand his rights; I demand his liberty without stint ... I demand that your brother be no longer trampled upon as your slave!'

Peterloo is virtually forgotten outside Manchester, partly due to the suppression of contemporary unbiased eye witness accounts, leaving an unbalanced written record. The magistrate in charge on the day was the Anglican churchman, the Rev William Hay, rewarded for his role in the massacre by immediately being given one of the richest livings in the country. His account of the happenings is the official one and in it, he praised the Anglican vicar Ethelston for the loudness with which he read the riot act. Suppressed reports say that Ethelston whispered the Riot Act out of a side window. Peterloo does not admit grey; purely black and white.

Possibly John Dalton himself can explain why the Manchester populus was less violent against dissenters than other towns in England. Non conformists permeated all levels of society in Manchester and the quiet devout Quaker John Dalton himself became revered in his lifetime. He was more than the equal of Priestley as a scientist and far less confrontational, if ever. Dalton was Professor of Mathematics and Natural Philosophy at the Manchester Academy during the six years that preceded the move to York, i.e. unlike Priestley in Warrington, he was hired to teach science.

Dalton wrote a letter to his cousin (an experienced meteorologist and instrument maker) Elihu Robinson, detailing the college and his life in it:

'Our Academy is a large and elegant building in the most elegant and retired street of the place; it consists of a front and two wings; the first floor of the front is the hall where most of the business is done; over it is a Library with about eight thousand volumes; over this are two rooms, one of which is mine; it is about eight yards by six, and above three high, has two windows and a fireplace, is handsomely papered, light, airy and retired; whether it is that philosophers like to approach as near to the Stars as they can, or that they choose to soar above the vulgar into a purer region of the atmosphere, I know not; but my apartment is full ten yards above the surface of the earth. One of the wings is occupied by Dr. Barnes's family, he is one of the tutors, and superintendent of the Seminary; the other is occupied by a family who manage the boarding and seventeen In-students with two tutors, each individual having a separate room, etc. Our Out-students from the town and neighbourhood at present amount to nine, which is as great a number as has been since the institution; they are of all religious professions; one friend's son from the town has entered since I came. The tutors are all Dissenters. Terms for In-students forty guineas per session (ten months): Out-students twelve guineas. Two tutors and the In-students all dine, etc., together in a room on purpose: we breakfast on tea at $8\frac{1}{2}$, dine at $1\frac{1}{2}$, drink tea at 5 and sup at $8\frac{1}{2}$; we fare as well as it is possible for anyone to do. At a small extra expense we can have any friend to dine, etc., with us in our respective rooms. My official department of tutor only requires my attendance upon the students twenty-one hours in the week; but I find it often expedient to prepare my lectures previously. There is in this town a large library, furnished with the best books in every art, science and language, which is open to all, gratis; when thou art apprised of this and such-like circumstances, when thou considerest me in my private apartment, undisturbed, having a good fire and a philosophical apparatus around me, thou wilt be able to form an opinion of whether I spend my time in slothful inactivity of body and mind. The watchword for my retiring to rest is - "past twelve o'clock, cloudy morning." '

In their report [47] of the 9th of August 1797 on the 'Academical Institution, or New College, at Manchester', the trustees expressed their satisfaction with Dalton, almost in the style of an end of term school report.

'In the province of Mathematics, Natural Philosophy and Chemistry, Mr Dalton has uniformly acquitted himself to the entire satisfaction of the Trustees; and has been happy in possessing the respect and attachment of his pupils. It is hoped and presumed, that he will continue, with zeal and ardour, his scientific exertions; and with the growing prosperity of the New College, he will enlarge his sphere of reputation and usefulness.'

Dalton was 30 years old when this report was written and had already published his first paper, on his own colour blindness. His lecture subjects included Mathematics and Geography, Natural Philosophy and Chemistry, theoretical and experimental. Dalton was far more than 'only a chemist' and the teaching of physics was in good hands, albeit for only three more years in this college. The claimed 'growing prosperity' of the college was at variance with the facts and painfully aware of the impending doom, Dalton resigned in 1800, to become a private lecturer and eminent researcher in physics and chemistry.

The 'large and elegant building' that Dalton referred to in his letter to Robinson was specially built for the college in Dawson Street (later Mosely Street) and is still marked as 'College Buildings' on the 1849 Ordnance Survey map (see Figure 1.21 on page 60), nearly half a century after the buildings were vacated as a teaching establishment.

The history of the Manchester Academy after it moved to York in 1803 has been described in great detail by John Seed [48]. In York, Seed says that it used the name 'Manchester College'. The three year course contained some science in each year, first year students being taught elements of geometry, algebra and trigonometry. The second and third year courses contained the geometry of solids, conic sections and spheres, advanced algebra, natural and experimental philosophy and chemistry.

The emphasis on science teaching to both lay and ministerial students was based not only on a desire to study 'God's creation of the Earth and cosmos as an act of reverence', but also to develop the mind and powers of reason. To the devout and socially elite, science was also seen as an integral part of an enlightened and cultured society. The scheduled lectures lasted from 7 am till 5 pm on weekdays and 7 am till 1 pm on Saturdays, with a break for lunch from 1 pm till 2 pm on weekdays. Extra lessons were fitted in at the end of the scheduled day. There were times when Wellbeloved taught the whole syllabus himself and his health suffered. Students also found it difficult to stay the course and drop-out rates were high. The subsequent financial strain on the sequence of Academies was one reason for their repeated failure. Poor strategy by the Academy governors also contributed. With a year in Oxford costing about £300, the academies were highly competitive in their early years, with fees and lodgings costing less than £50 a year. But the York Academy raised their fees alone to over £100 per annum and discovered that too few Northern parents of prospective students could afford it. Perhaps no other college has ever wrestled with its own name to the extent that this one has. In his book of 1932, V D Davis describes [32] how the name of the college evolved:

> ' 'The Manchester Academy' was the name first adopted, in succession to Warrington. In a circular of 1797, it is called the 'Academical Institution or New College at Manchester,' and George Walker, on the title-page of a volume of his Essays in 1809, is designated 'late Professor of Theology in the New College and President of the Philosophical and Literary Society, Manchester.' A College document of 1805 bears the title 'The New College, Removed to York;' others of 1807, 1810, etc., 'Manchester New College, Removed to York.' Letters of John James Tayler, in 1816 and 1818, are dated as from 'Manchester New College, York'; but during the York period the prevailing use was simply 'Manchester College, York.' On the return to Manchester in 1840 the regular use of 'Manchester New College' began, and continued throughout the London period. It was after the opening of the College buildings at Oxford, in October, 1893, in the near neighbourhood of the ancient University foundation of New College, that the 'New' was eliminated from the name of 'Manchester College' '.

The enduring legacy from this Academy or 'New College' is that it was responsible for attracting John Dalton to Manchester in 1793, where he stayed for the remaining 51 years of his life, long after the college left without him.

1793–1844: John Dalton

The years 1793 to 1844 represent the period from when John Dalton arrived in Manchester to take up his post as tutor at the 'New College' until his death. He is the first scientist on record to start a programme of scientific research into physics and chemistry in Manchester and to carry it out at the highest level.

Brief biography

Dalton bequeathed 'all his philosophical, scientific, and literary manuscripts and correspondence' to William Charles Henry and three other executors. With Henry out of the country on Dalton's death in 1844, the relevant manuscripts 'fell into the hands' of executor Peter Clare, who declined to hand them over to enable Henry to write a biography. Thwarted for ten years, Henry completed the task in 1854 upon the death of Clare. I have used this biography [49] as source material on Dalton, rather than any later 2nd hand version.

Figure 1.13:

John Dalton 1834. Derived from a stack of various monochrome engravings and mezzotints.

Colour processing © 2017 Robin Marshall.

Dalton became closely tied to the Lit. and Phil. when he resigned from the New College, just as the college itself had been entwined with the society during its brief stay in town. After holding its early meetings in an inn and the Assembly Coffee House, the society met in a large room adjoining the Cross Street Chapel before moving to their long-term home in 36 George Street. Dalton was provided with a room in which to study and do research and he lodged for many years in the house of his friend Rev W Johns at 10 George Street, a few doors away.

Henry's biography encapsulated the essential lifestyle and financial needs of Dalton in a few sentences [49]:

'During the winter of 1787, he delivered, at Kendal, a course of twelve lectures on natural philosophy, comprising mechanics, optics, pneumatics, astronomy, and the use of the globes; a course which he repeated in 1791, with the addition of a lecture on fire. His terms of admittance to his second course were exceedingly moderate, being sixpence each lecture, or five shillings the whole, – half the sums announced for the earlier course. Possibly he may not have found in the Kendal of the last century, a sufficiently large audience at the higher fee. From this period it became a part of his regular occupations and an important source of his slender revenues, to deliver lectures in Manchester and elsewhere. In the spring of 1793, he was invited, mainly in consequence of Mr Gough's favourable recommendation, to join a college established in Manchester, in the year 1786, by a body of Protestant Dissenters, as tutor in the department of mathematics and natural philosophy. The terms proposed and accepted by Dalton were, that he should receive three guineas per session from each student attending his lectures, with the condition that the sum shall not fall below £80/- per session of ten months. Commons and rooms in the college were allotted him at £27. 10s. per session. He resigned this appointment after a period of six years; but continued to reside in Manchester during the whole of his subsequent life.'

Henry describes Dalton as being of medium height, 5ft 7in, which would now be regarded as diminutive for a male. He had a deep, gruff voice. Let me now not re-invent the chemist John Dalton as a physicist, he was clearly both, but take a look at some of the research and discoveries he published, which are clearly more physics than chemistry. His early work on gases, partial pressure and establishing the gas law is often overlooked in favour of the atomic theory. But if Nobel prizes had been awarded 100 years earlier, it is quite likely that he would have shared one with Gay-Lussac for establishing one half of the gas law.

The Gas law (revisited)

Dalton read four Experimental Essays at the Friday meetings of the Lit. and Phil. on the 2nd, 16th and 30th of October 1801 [50].They eventually appeared in print in volume five of the Society's Memoirs, which were published in 1802. The subject matter is clearly physics although the distinction from chemistry would have been less distinct in 1801.

The first essay dealt with the constitution of mixed gases, especially the atmosphere. 24 years had elapsed since Lavoisier had announced that the atmosphere was not a unified fluid, but contained nitrogen and oxygen. Dalton first dismissed all hypotheses where the particles of the gases attracted, repelled or chemically interacted with each other. Some of the reasons for rejecting the hypotheses were invalid, since it was not known why the heavier oxygen did not form a layer beneath nitrogen.

Using simple arguments and a few mathematical formulae that even aspiring officers at the Woolwich Academy would have found trivial, Dalton proceeded to produce his law of partial pressures. Despite the derivation's lack of rigour, the law has endured as a very good approximation to the truth. Dalton finished by reminding the audience that in his *Meteorological Essays*, printed in the previous volume of the Memoirs, he had shown, after being inspired by the 'aurora borialis', that a fluid (gas or vapour) possessing magnetic properties must hold a place in the higher regions of the atmosphere. The fluid, Dalton asserted, must be possessed 'of a *ferruginous* quality; but it will probably ever be beyond the reach of philosophical research to ascertain the nature of so subtle and distant a fluid.' Until 1820, when Hans Christian Oersted discovered that a current-carrying wire would produce a magnetic field, magnetism was exclusively associated with ferrous-ferric lodestones and the earth. Dalton certainly did not think that there was a layer of iron, but unlike Kepler, who was able to incorporate a body (Saturn) 1.433 billion km from the Sun into his desktop calculations, Dalton banished the iron-like layer to the realms of unmeasurable. But his vision of (then) unknown electrons curving in the earth's magnetic field was profound.

The second essay concerned the force of steam or vapour from water and various other liquids. The extant terminology used by Dalton was that *steam* or a *vapour* could be liquefied whereas a *gas* could not, although as he pointed out, he believed that all gases could be liquefied given a sufficiently low temperature or high pressure. His aim was to measure and study the relation between pressure and temperature for the vapours of water, alcohol, ether, liquid ammonia, calcium chloride (solution), mercury and sulphuric acid. After a plethora of tables and results, he noted that:

> 'The expansion of all elastic fluids, it seems probable, is alike or nearly so, in like circumstances; 1000 parts of any elastic fluid expands nearly in a uniform manner into 1370 or 1380 parts by 180° of heat.'

But before that, essay three was presented, on evaporation. This was a lengthy qualitative account of the evaporation rates of various liquids under various conditions, and the fact that the experiments were carried out under uncontrolled domestic conditions, using open windows and breezes to change the evaporation rates diminished their worth.

The fourth essay was essentially a continuation of the second, but this time applied to non-liquefiable gases. The main conclusion of the paper was that all the gases he investigated (hydrogen, oxygen, carbon dioxide and nitrogen) expanded by the same amount between the fixed ice and steam points. Gay-Lussac came to the same conclusion in his later publication [51]. There has been much debate about the priority between Dalton and Gay-Lussac, even in recent times. But both publications are dated and Dalton's presentation of 1801 was several months ahead of Gay-Lussac's in 1802, time enough for news to get from Manchester to France, although that is mere supposition.

How then did the simple gas law, stating that volume is directly proportional to temperature come to be erroneously known as Gay-Lussac's law and even more erroneously, as Charles' law for some time? In his paper, Gay-Lussac credited unpublished work by Charles carried out several decades earlier. But whatever it was that Charles did, or actually reported, is not on the record and Gay-Lussac, with measurements at only two temperatures was unable to determine the functional relationship. Dalton made measurements at four temperatures, including the results from Essay II.

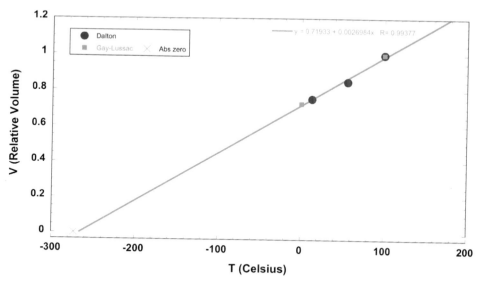

Figure 1.14: *The combined data of Dalton and Gay-Lussac. The linear fit passes through zero volume at –267 C. The × marks the internationally agreed value of –273.15 C.*

I have put the combined data from Dalton and Gay-Lussac into a single plot, as shown in Figure 1.14 and fitted a straight line to them. The straight line is an extremely good fit, far better than alternative hypotheses of quadratic or cubic. Even though it was not normal to publish data in graph form at that time, nor fit curves to data, but why did Dalton not pounce or pronounce on the linearity? It was the simplest hypothesis and not at odds with the data. In truth, Dalton trusted the accuracy of his thermometers too much. He found that the expansion to the half way point was 167 parts compared to 158 for the second half. Even towards the end of the 19th century, the vagaries of mercury thermometers between the fixed points was a matter of endless discussion and investigation.

Dalton changed his mind about the temperature dependence of volume within five years, but still got it wrong. He presented a detailed table of the mercury and gas temperature scales in a compilation of lectures [52] presented in Glasgow and Edinburgh in 1807, declaring that the volume increased in geometric progression with temperature. The half way split of 167:158 (decreasing expansion) in his essay now became 152:179 (increasing expansion). These two conflicting results were never addressed or remarked on by Dalton, who surely cannot have failed to notice that the results vacillated around the linear function, which was manifestly the most plausible and appealing, although he never said it. Thus we find that neither Gay-Lussac nor Dalton separately claimed or justified the linear relation between pressure and volume.

The linear fit that is shown in Figure 1.14 on the previous page, extrapolates to −267 C at zero volume. These data do not prove the existence of an absolute zero, but say that if the data are extrapolated several factors below their span, through and beyond a known physical discontinuity where vapours of water and ether liquefy, zero volume is reached at a certain temperature, at which the volume of the already non-existent vapour might be considered to be meaningless anyway.

Dalton could have achieved this result himself but the serious obstacle of his faith in the caloric theory got in the way. He believed that the particles of a gas occupied no space and it was the substance of heat, 'caloric' that filled the volume, entering and leaving the gas as its heat was increased or decreased. He then adapted an extraordinary procedure, for which there appears to be no plausible reason apart from the fact that his result agreed with one of his heroes, the illustrious Irish implanted Scotch descended physician and chemist Adair Crawford, a devout but not evangelical phlogistonist. Crawford had determined absolute zero to be at 1532 F below freezing (−869 C) by a method too long and wrong to be included in this book. Apart from this, Crawford's work on the thermodynamics of living creatures preceded the establishment of thermodynamics itself and many of his results were taken over without accreditation by Joule's competitor, Julius Mayer, for priority in the discovery of the mechanical equivalent of heat. To obtain his value for the absolute zero of temperature, Dalton first advanced the hypothesis that the repulsive force of each constituent particle of the gas was exactly proportional to the heat of that particle, which he correctly set proportional to temperature, even realising that the temperature should be measured from the point of 'absolute privation' or 'absolute cold', which was the complete absence of heat. The space between particles was filled with caloric, whose volume shrank with decreasing temperature (heat). His big mistake was to argue that the heat would be zero when the force was zero and this would happen when the range of the force, which he not unreasonably set proportional to the cube root of the volume was reduced to zero. Dalton's cubic extrapolation gave 1547 F below freezing (-877 C) and he could not refrain from making himself a hostage to future better physics knowledge by remarking 'So near a coincidence is certainly more than fortuitous.'

Figure 1.15:

Derived from an 1838 monochrome Mezzotint of William Henry by James Turner, from various engravings of oil paintings, long since lost.

Colour processing © *2017 Robin Marshall.*

If Dalton had paid more attention to the paper that followed his in the Memoirs, written by William Henry (see Figure 1.15), the father of his eventual biographer, William Charles Henry, he would have realised that great minds like Count Rumford and Humphrey Davy had taken the bold step of assigning the cause of heat to motion among the particles. He could then have made a linear extrapolation to zero motion and assigned this to zero volume as well, a far more plausible assumption than the range of mythical, calorific forces decreasing like the third power of temperature. It might be too much to expect that Dalton, a dedicated calorist, should have listened to his friend Henry's paper, comparing the caloric and dynamic theories of heat and then disagreed with Henry's (wrong) conclusion. Accepting that a useful way to compare the two theories was to take the path of *reductio ad absurdum*, Henry proceeded to reduce the caloric theory to absurdity, before declaring it was correct. One can imagine Dalton on the front row, nodding his head in agreement. Henry made his reputation as a chemist and even more so as a manufacturer of fizzy drinks. This is the only foray he made into physics that I shall mention.

Figure 1.16: *The montage of Men of Science 1807.*

Both John Dalton and William Henry were included in Sir John Gilbert's montage of 'The Men of Science Living in 1807–8', which is shown in Figure 1.16. Dalton is at top table, pensive, and Henry stands to the left of Dalton as we look, dark hair and dark jacket collar. A rather bland Humphry Davy stands directly behind Dalton. James Watt is also at top table, sitting in the grand chair on the right. Standing behind him is Count Rumford who effectively disproved the caloric theory by continuously boring the barrels of cannons, producing continuous heat

54

and swarf with the same heat capacity that it possessed when part of the cannon. Dalton and Henry should have persevered with Rumford's interminably long papers.

Funeral

Dalton's death on the 27th of July 1844 precipitated an immense reaction, not only in Manchester, which had only just been incorporated as a city six years previously, but also throughout the nation. His death, his lying-in-state in the (old) Town Hall in King Street on Saturday the 10th of August and his funeral on Monday the 12th of August were given extensive coverage in the media. The *Illustrated London News* (ILN) in particular published engravings of sketches made in the Town Hall, where Dalton's mahogany coffin was laid out.

The scene inside the Town Hall, illuminated by eight large candles, was described in detail and is shown in the engraving reproduced in Figure 1.17. Someone, probably several tellers, carefully counted the number of persons entering the Town Hall to pay respects and they recorded an initial surge of 110 persons a minute for one and a half hours, whereafter the number varied from 90 to 100 per minute and after eight hours, so the ILN reported, *forty thousand* persons had passed through the room.

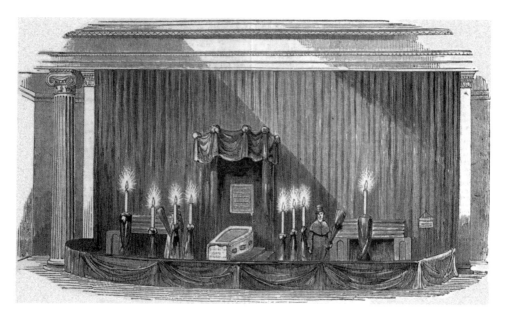

Figure 1.17: *John Dalton, his lying-in-state in Manchester Town Hall.*

The hearse arrived at the Town Hall at 10 am on the Monday morning, where the procession was organised. The cortège eventually arrived at Ardwick Cemetery, 1¼ miles away at twenty minutes past midday, after passing through the streets tightly thronged with crowds. The scene was faithfully recorded as the ILN engraving reproduced in Figure 1.18 shows. Everyone dressed up in their finery to salute Manchester's first true physicist as the parade went from the old Town Hall in King Street, past the Lit. and Phil. premises in George Street, near his house in nearby Faulkner Street, in what is now Chinatown. No physicist before Dalton, and no one since, has received such unbridled adulation.

Figure 1.18: *The funeral procession of John Dalton, on its way to Ardwick Cemetery.*

1824: The Manchester Mechanics' Institution

Historically, Mechanics' Institutes were educational establishments formed to provide adult education, particularly in technical subjects, to working men. As such, they were often funded by local industrialists on the grounds that they would ultimately benefit from having more knowledgeable and skilled employees. Such philanthropy was shown by, among others, Robert Stephenson, James Nasmyth and Joseph Whitworth. The Mechanics' Institutes were also used as libraries for the adult working class, and provided them with an alternative pastime to gambling and drinking in pubs.

The first Mechanics' Institute was incorporated in Glasgow in November 1823, built on the foundations of an enterprise started at the turn of the previous century by George Birkbeck. Under the auspices of the Andersonian University (established in 1796, now the University of Strathclyde), Birkbeck had first set up free lectures on arts, science and technical subjects in 1800. These classes continued after he moved to London in 1804, when their organisation was incorporated as the 'Mechanics' Institute'. The London Mechanics' Institute, later Birkbeck College, where Manchester's future colossus Patrick Blackett would be a professor, followed in December 1823, with the Mechanics' Institutes in Ipswich and Manchester in 1824. The fact that the lectures were free probably explains why the institutes prospered, compared to their fee-paying competitors.

The Manchester Mechanics' Institute (or Institution as it was originally known) was set up in 1824 by means of local private initiative and funds with the intention of teaching the sons of artisans the basic principles of science via part-time courses, thus allowing them to continue working. The science was chemistry, mathematics and some natural philosophy (physics) in the form of mechanics. The founding meeting was convened by George William Wood on the 7th of April 1824 with Sir Benjamin Heywood, a prosperous banker and the Institution's first President in the chair. Also present was Robert Hyde Greg, a cotton mill owner and soon-to-be MP. Peter Ewart, millwright, engineer and physicist, whom we shall soon meet again as a powerful and crucial link between John Smeaton and James Joule was also there, along with John Dalton, who became a Vice President of the Institution in 1840. Richard Roberts, inventor of various machine tools, the builder David Bellhouse and William Henry, now flourishing as an aerated drinks manufacturer were inevitably present. William Fairbairn, more of whom later, the versatile and illustrious engineer, whose name is associated with water wheels and machinery of every kind as well as the Britannia tubular bridge, was elected its first Secretary. An action committee was then elected to realise the planned institution, including Wood, Fairbairn, Heywood, Roberts and one of Joule's future tutors, John Davies. The Institution soon opened in 1825 with Heywood at the helm. The list of energetic, eminent and wealthy supporters and sponsors that drove the creation of the Institution is unparalleled in the history of

Manchester colleges, academies and institutions and is one of the reasons why the Institution survived and thrived, when so many before it had crumbled. There were no religious cliques waiting to cause schisms and rifts and the education it provided was almost free.

Figure 1.19: *The Cross Street Chapel shown here in about 1835, where the Mechanics' Institution, the forerunner of The University of Manchester, held its first science classes from 1825 to 1827.*

Members of the Lit. and Phil. were closely involved in setting up the Institution and for the first two years of its existence, classes were held in the Cross Street Chapel which the 'Lit. and Phil.' themselves had used from their foundation in 1781 until they acquired their own premises on George Street in 1799. The chapel, shown in Figure 1.19 had been built in 1694 and surviving the riots of 1715 with a minor rebuild, stood until Christmas 1940 when it was destroyed in the Manchester Blitz.

The chapel, purporting to be as it was in 1835, was drawn by Fred W Goolden and lest we should think that he set up his easel in Queen Street, looking out across Cross Street, the drawing is attributed to 1910. Either some earlier image was copied by Fred, or he drew the existing building as he saw it and used his imagination. Putting aside this small detail, the chapel in 1835, the erstwhile premises of the Manchester

Literary and Philosophical founded in 1781 and the first home of the Manchester Mechanics' Institution, is presented in Figure 1.19, a superb drawing.

Figure 1.20: *The Mechanics' Institute Building in Cooper Street near what is now St Peter's Square in 1827.*

In 1827, the Institution moved to its own premises (see Figure 1.20) in Cooper Street, very close to the current Town Hall. The detailed Ordnance Survey map, published in 1849 (but surveyed in 1844) showed the internal structure of large buildings and in the case of the Mechanics' Institution, the large lecture theatre can be appreciated on this sixty inch to the mile map, (see Figure 1.21 on the next page).

The 'College Buildings' on the Ordnance Survey map were those built for the Manchester New College when it was set up in 1786 and they were turned into a gentleman's club after the College departed in 1803. No drawing or engraving of the buildings appears to have survived and they were demolished along with Cooper Street and that part of Mosely Street when the area was reshaped and redeveloped for the Town Hall in the 1870s.

There was an early minor perturbation in 1829, when the self taught radical politician and Swedenborgian preacher Rowland Detrosier, rightly believing the Mechanics' Institution to be undemocratic, formed a 'New Mechanics' Institution' in Poole Street, a move that had a serious effect on the recruitment and finances of the original establishment. Detrosier himself attracted a significant number of subscribers, as the running sequence number on the subscription ticket of William Ashton, shown in Figure 1.22 indicates.

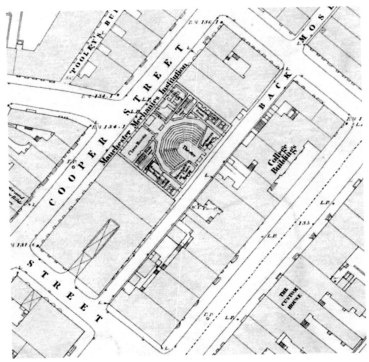

Figure 1.21:

A section from the 1849 Ordnance Survey showing the Manchester Mechanics' Institution and the site of the former Manchester New College ('College Buildings'). College lecturers would have swapped cigars in Back Mosely Street.

Figure 1.22:

William Ashton's subscriber ticket to the New Mechanics' Institution, signed by the Institution's founder Rowland Detrosier.

Image: Courtesy, The Lilly Library, Indiana University, Bloomington, Indiana.

Ashton eventually emigrated to the USA, taking his ticket with him. It is possible that in the past, historians sympathetic to the original Institution have understated the depth of this fracture, which nearly brought about the early downfall of the Institution.

In any case, subscriptions and memberships of the original Institution fell to an all-time low in 1830–31 and only the gradual opening up of the board to election by the members rectified the situation. Board members were normally appointed partly on their willingness to pay an annual subscription. Eventually, Detrosier's break-away establishment rejoined the original institute and growth resumed.

By 1840 the Institution was thriving, with 1,000 subscribers and a library of some 5,500 books. However, the increased popularity had been achieved at the expense of science education, and an increasing number of non-scientific lectures were occupying its programmes. Just as the Institution began to prosper, the wandering Warrington, Manchester, York Academy returned, to be followed three years later by a Congregational college relocating to Manchester from Blackburn. The second transient passage of the most itinerant academy did nothing to halt the relentless progress of the Mechanics' Institution, which found a resonance with what Manchester had now become. After the Foundation of the Owens College in 1851, The Mechanics' Institution became increasingly connected with Owens and then with the subsequent Victoria University of Leeds, Liverpool and Manchester colleges. It eventually became part of the University of Manchester as a Faculty of Technology in 1905, and then a University in its own right, UMIST in 1994, after which it became an equal partner in the 2005 merger between the University of Manchester and UMIST.

1829–1836: Reform, but still no University

The last decade of the 18th century was not a pleasant one for a country which had been calling itself since 1707, the 'United Kingdom' and riots prevailed in many cities. Manchester, the hub of the industrial revolution, was legally still a feudal baronial village and hence, not a city. It was possible to count up 130 rotten boroughs whose summed population equalled that of Manchester, each returning two Members of Parliament, 260 MPs of them open to bribery and not surprisingly succumbing to it. Yorkshire with a population of one million had two MPs, the same as Old Sarum, whose total population lived in three houses. Manchester with a population of 60,000 had none, despite contributing more via taxation and duties than any other city to the nation's wealth. In 1791, Thomas Paine had wondered if there were any principle in these things. Small wonder there were riots. Small wonder those in power sought to suppress them with violence. Small wonder that Paine's adopted colony across the Atlantic threw off the yoke.

In this climate, there was even less of a chance than there had been in 1641 of creating a university in Manchester. It was not Manchester specifically that the Oxford and Cambridge 'now universities' were against. But like the fading leader of the wolf pack, whose ancient teeth clang against each other instead of sinking into the flesh of the prey, the days of the old universities' monopoly were numbered. They tried, almost as an auto-genetic response, to prevent the creation of London University but inevitably failed. But their time had run out. Thereafter, the creation of universities elsewhere became inevitable, although still requiring patience and effort.

Even so, in Victorian England, it was not easy to throw off the yoke of Oxford and Cambridge's exclusivity and three quarters of the 19th century were consumed by repeated attempts to have a university in Manchester. Political decisions in Parliament were still being made by cliques like the two MPs for Old Sarum in Wiltshire, who had been elected by seven voters. They were probably in alliance with the two MPs elected by the seven voters of Gatton in Surrey. Even at the 1831 election, 152 of the 406 'elected' Members of Parliament were hand picked by fewer than 100 voters. The 1832 Reform Act, despite being opposed by those in power, and being the cause of several political resignations, seized the tip of the nettle and disenfranchised the corrupt boroughs of Old Sarum and Gatton and another 55 like them. Time enough, one might have thought, for a university in the North. Not so. The next sections deal with a few further failed attempts, one of which, let truth be told, was self inflicted.

1829: A University at the Royal Manchester Institution

The Royal Manchester Institution had been established in 1823 with the intention of bringing more culture and taste to the increasingly industrialised baronial village. It occupied what is now a Grade I listed building housing the city's art gallery. By attracting the patronage of George IV it acquired the designation 'Royal' and the King donated more than his name by sending plaster casts of the Elgin Marbles, which he 'owned', to decorate the entrance. The Manchester Mechanics' Institution aimed to educate the artisan, a move regarded with some trepidation by the mill owners who feared their own sons might become less well educated than their workers. Suppressing the Institution would have been un-Mancunian and so they sought to combat the threat by setting up a refined Institution of their own.

'Thus, by 1824, Manchester had both a 'Royal' and a 'Mechanics'' Institution. Only one Institution survived. But while it lasted, the Royal Institution could not be faulted on the quality of its lectures. At a time when physics was establishing itself in England as a subject, the Royal Manchester Institution offered three courses on astronomy by John Nichol and three on electricity by William Sturgeon who was without question, the Mr Electricity of his day. The Institution illuminated its lecturers with the title 'Honorary Professor'.

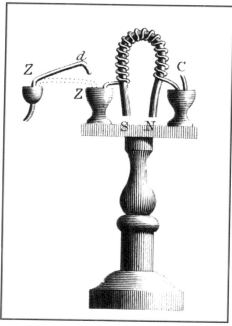

Figure 1.23: *Left: William Sturgeon. Right: His first electro-magnet, invented in 1824. The image has been created from an engraving of a lost oil painting.*
Colour processing © 2017 Robin Marshall.

Sturgeon had seen little future in being apprenticed to his father, at best as a cobbler and at worst, as an idle poacher. He therefore joined the army in 1802, at a time when the British Army had much to do. His interest in science was ignited by watching a thunderstorm in Newfoundland whilst fighting the French. He was unable to obtain a satisfactory explanation for lightning and set about learning about electricity, making many fundamental contributions to the new science. He was a genuine physicist, a lecturer on his chosen speciality and publisher of the first journal on the subject, the *Annals of Electricity*.

Sturgeon is the acknowledged inventor of the electromagnet (see references [53] and [54]) and his first working version, published in 1824 [55], can be seen in Figure 1.23 on the previous page and also on the cover of this book. This 1824 paper marked and recorded the award of a large silver medal and 30 guineas by the *Society, Instituted at London, for the Encouragement of Arts, Manufactures, and Commerce*. A more glittering career might have been expected of him, but he was never elected FRS and he had a tendency to drift from one failed project to another, failing by virtue of a brusque personal demeanour and an inability to foster a working relationship with those whom he disliked. He was also involved in two further short-lived institutional enterprises.

'The Royal Victorian Gallery for the Encouragement and Illustration of Practical Science', essentially a science museum, opened in 1840 in the Dining Room of the Royal Exchange buildings in Manchester, with Sturgeon as the gallery superintendent. James Joule gave his first public lecture there in 1841 and after the gallery folded in 1842, Joule remarked [54] in the Lit. and Phil. Memoirs:

'... the indifference to pursuits of an elevated character which too frequently marks wealthy trading communities destroyed this, as it has many other useful institutions.'

Isolated but undeterred, Sturgeon then announced 'The Manchester Institute of Natural and Experimental Science' in January 1843. With subscriptions pitched at the same level as the Royal Victorian Gallery, which had failed to attract enough support, it was no wonder that this Institute ended before it had barely started. In the meantime, he continued as before, lecturing occasionally at the Royal Institution and publishing his 'Annals', which remain one of his lasting contributions to Manchester physics. They served as a launch pad for Joule's career and later, when Joule suffered like Dalton had with questions of priority, Sturgeon reprinted relevant papers from earlier out-of-print editions to establish Joule's claim. It is an interesting counterpoint to modern trends that sometimes, the impact of the science being published outweighs the impact of the journal in which it is published. He eventually became an independent, roaming lecturer, taking his demonstration apparatus with him in a cart, although not on the successful scale of Adam Walker, 50 years previously. His tombstone in the churchyard of St Mary's, Prestwich was inscribed 'William Sturgeon – The Electrician'.

Figure 1.24: *The Royal Institution Building in Mosely Street, now the City's Art Gallery, another potential home for Manchester University in the 19th century.*

Photo: Robin Marshall.

During this period covering the first quarter of the 19th century, physics education in Manchester was stuttering, with little continuity. On the research side, however, Dalton and then Joule placed Manchester on the world stage.

The Royal Institution building (see Figure 1.24) was and still is, magnificent by any standards. It was the first large commission for architect Sir Charles Barry FRS who went on to design the new Palace of Westminster after the 1834 fire.

In 1829, William Whatton, a governor of the Royal Institution wrote to fellow governors advocating the establishment of a university [56]. Pointing out that almost two million people lived within a radius of 20 miles and that a leading member of the administration (government) had pledged that Manchester shall send its own Members to Parliament at the 'first favourable opportunity', he advanced further arguments in favour of a good and efficient system of education. Manchester had superabundant wealth, the Literary and Philosophical Society, The Natural History Society, the Royal Manchester Institution, the Mechanics' Institution, the Botanic Gardens and two wealthy schools.

Throwing open the doors of literature and science, he would convert the (Royal) Institution into a University, where middle class youths could study literature, science and the arts. The closest that Whatton got to physics was to mention mechanics, including mathematics. The splendid building in Mosely Street would be admirably adapted for the purpose, he declared, and the future Manchester merchants would benefit from a knowledge of foreign languages and the calico printers would derive much consequence from learning about the chemistry.

Whatton gave no reasons for including or excluding physics, or even engineering. His noble proposals for higher education in the city go some way towards ameliorating the appalling answers he gave to a House of Lords Committee on the 25th of May 1818, judged by the standards of any age. As a surgeon examining children as young as eight who worked a twelve hour day in the factories of Manchester, he claimed never to have observed any ill effects from the factory work. 'Even a child of delicate constitution,' he said, 'would not be injured by labour so moderate, that could scarcely be called labour.'

It is easy, but probably justified, to think that Whatton did not know the meaning of a 'hard days labour', as opposed to a 'hard days work', still an issue of the colour of the collar in the 21st century, but now mercifully, no longer an issue of sending 8-year olds into factories. After a second pamphlet three months later where he replied to objections to his scheme, nothing was ever heard again of Whatton's initiative and by the time Owens College was founded, the Royal Institution had essentially served its brief but ephemeral purpose of science instruction for local males. It is usually unfair to judge those who fail modern standards, judged by modern standards. But even judged by the standards of 1818, Whatton fails; one can only ponder on the size of his retainer from the mill owners.

1834: The Word 'Physics' enters The English Vocabulary

In 1834, the British Association followed the French lead and formally recognised physics as a subject in its own right and by 1840, the word *physicist* was coined, it being recognised that the French word *physicien* was unacceptably confusing. The word *scientist*, not in common use for general purposes was also considered and rejected. The term 'Natural Philosophy' had been used previously and continued in parallel as a title for physics, especially in Scotland. The 'Department of Natural Philosophy' at Glasgow University was not changed to 'Department of Physics and Astronomy' until 1986.

Manchester University used the designation 'Natural Philosophy' from the inception of Owens' College in 1851 until the appointment of Balfour Stewart and Thomas Core as professors of physics in 1870.

1836: A University in Piccadilly

William Fairbairn was the penultimate unsuccessful proposer of a university for Manchester, which he mentioned in his grand 1836 proposal [57] to improve the town of Manchester by blending the chaste and beautiful with the ornamental and useful. As noted by Thompson [1], it also blended the costly. A sweeping crescent was envisioned (see Figure 1.25) along the North-West side of Piccadilly facing the Infirmary, the front of which was to be adorned with a statue of Hygeia, the goddess of health, cleanliness and sanitation. She would be gazed upon beaconly by strategically placed statues of James Watt, Richard Arkwright and the Duke of Bridgewater. The new curved Exchange, which would double as the University, was to be located at the corner of Market Street and Mosely Street.

Figure 1.25: *Fairbairn's vision for a university (the curved building in the right foreground) on the corner of Market Street and Mosely Street (which enters on the right, behind the university). The centre foreground statue overlooks Piccadilly with the Infirmary on the right.* *Image from reference [57].*

The multi-talented Fairbairn had enlisted his multi-talented friend, architect, engineer, astro-physicist and illustrator James Nasmyth to do the designs and drawings. Fairbairn was fulsome in his praise of Nasmyth,

67

who deftly devised aesthetic chimneys for dispersing the smoke from the coal powered steam engines. (See Figure 1.26 for portraits of Fairbairn and Nasmyth.) There were no chimneys in Nasmyth's drawing, which also lacked the allegorical figure of Mercury at the apex of the university building for which error, Fairbairn apologised profusely in his treatise, it being presumably cheaper to apologise than to re-engrave the plates. He did not apologise for omitting the essential chimneys. Mercury was part of the plan to 'effect so important a *desideratum*'.

Figure 1.26: *Left: William Fairbairn, ca 1852. Right: James Nasmyth, ca 1844.*
Colour processing and graphics © 2017 Robin Marshall.

Fairbairn's prose was as ornamental as the architecture. Appealing to the impressions made on posterity by Egyptians, Assyrians, Persians, Greeks and Romans, conceding that he might be proposing a Utopian scheme, he then enlisted Chaldeans, Phoenicians and Tuscans to his aid. Chrysippus, Cleanthus, Plato and Pythagoras were also enjoined to give the statues justification and Edinburgh was cited as an example of how not to adapt a city to the tastes of modern society. The notion of a university within Fairbairn's scheme for Manchester is presented almost as an afterthought. But he was, as noted above, a founder and the first secretary of the Mechanics' Institution, to which the present University of Manchester traces its origin. Fairbairn had national stature, power and influence, which he used wisely and philanthropically. But his design for a university in Manchester never went beyond paper.

The original mezzotint of William Fairbairn from which the colour image was derived (Figure 1.26) was engraved by Thomas Oldham Barlow from a painting by Philip Westcott and published by the Manchester based fine arts dealer and publisher Thomas Agnew in 1852. Fifty copies were released. The superb photograph of Nasmyth, used as basis for Figure 1.26 was taken in 1844 by the newly founded Scottish studio of Hill and Adamson.

We shall shortly see how John Owens set about making a university by a different route. He wrote his will in such a way that the bulk of his inheritance went to founding a college that then became a university. The text of his will and the choice of his Trustees spoke volumes. Fairbairn's more voluminous 49 pages on his vision for Manchester simply make for entertaining reading, nearly two centuries on. His dedication to Manchester was underlined when he chose the title (First) Baron of Ardwick, when he was elevated to the peerage.

1836: Harry Longueville Jones' University

William Fairbairn's charmingly buoyant vision for a university in Manchester, although without substance from the educational point of view, was given strong local support and the required substance soon followed from an unexpected source in the form of a rigorous paper [58], *Plan of a University for the town of Manchester*, read before the Manchester Statistical Society by Rev Harry Longueville Jones in the summer of 1836. Jones, a Fellow of Magdalene College, was an educationalist of missionary zeal who came to dislike the way Oxbridge operated with the sort of intense passion that is reserved for those who have passed through the 'system' they eventually come to despise. He was appointed Inspector of Church Schools in Wales in 1838, which dissipated his energy into areas that Oxbridge did not need to worry about. But before then, his presentation in Manchester was so much admired that fellow educationalist, James Heywood, son of the first President of the Mechanics' Institution and himself President of the Manchester Athenaeum, paid for Jones' paper to be published as a pamphlet out of his own pocket and discussions followed on how to establish the scheme for a college in Manchester.

Welsh born and Oxford educated Jones was fulsome in his view of Manchester [58]:

> 'In all directions the circle of Manchester is full of life and intelligence, manufactures of every kind occupy the inhabitants of the towns; the movement of money is immense; commercial activity is carried to an extraordinary pitch; mechanical ingenuity receives there daily new developments; the minds of men are in a state of electric communication of ideas; their political sentiments indicate the restless vigour of a rising and sturdy people; their religious opinions are full of fervour and piety. Yet one thing still is wanting - the vast population of South Lancashire wants a centre of intelligence and moral improvement; it requires one if not two "seminaries of sound learning and religious education".'

His natural desire as a Reverend to make it a religious university, in Manchester, which would have condemned it to rapid failure, can be overlooked, given the far wider scope and quality of his proposal.

A preparatory meeting was held at the York Hotel on the 10th of November 1836 and 87 prominent locals, the Boroughreeve and the Constables of Manchester were invited to attend and form a committee. Richard Cobden, John Davies (lecturer and tutor to Joules), chemist Thomas Henry, the ubiquitous William Fairbairn and a young physics professor from the Manchester New College, Montagu Lyon Phillips were present. Whether Montagu Phillips had thought that physics would receive peer treatment in the proposed new college, we shall never know. But if he had, his hopes were quickly dashed with the declaration that no more than six professors should be created in the *Faculty of Arts* (sic), to cover the subjects of:

1: Mathematics, pure and mixed.
2: Chemistry with allied branches of physical science.
3: Natural History, with physical geography.
4: Classical literature.
5: English literature.
6: History with economical and political history.

Still in the age and legacy of Dalton, physics was a mere subset of chemistry. The meeting proposed and decided that classes would be held in the Royal Institution in Mosely Street. Schools of design and modern European languages were put off for the future and the committee decided to go away and think about a medical faculty and a law school.

By 1837, it had been decided to include a medical faculty, but there being two rival medical schools in Manchester, the failure, if not total inability of these two factions to agree a common path, was a contributory cause to the demise of the whole venture. The ego-driven squabbling between the various medical factions in Manchester during the first half of the 19th century deserves a book of its own and was contemporarily surpassed only by internecine schisms at the New College (see below) which managed to pick an 'intellectual' fight with itself, as many religious based edifices do.

This new initiative failed before it even got started. Thompson [1] dryly remarked:

> 'Jealousies had unfortunately arisen between the (*medical*) schools, and although these seemed to have been composed by the energy of the committee, they ultimately rendered it impossible to prepare a course of lectures by joint action for the season of 1837–8, and before another year had passed the college scheme had been abandoned.'

Thus, after 1838, nothing more was heard of this venture, although when Owens College was founded in 1851, the proposed governance and teaching infrastructure of Harry Longueville Jones' university were noted to be so substantial and professionally worked, that they were taken over as a template, together with a large number of its supporters.

Six years later in 1844, John Owens died, prematurely by any measure at the age of 55 and a mere seven years after his death, Owens College was founded. It was a close run thing, as we shall see in the next chapter. The latest college had only been in existence for seven years when the *Manchester Guardian*, in its leading article of the 9th of July 1858, distinctly pronounced the college a failure. This sounds like the same old story, except this time, the college came through the crisis, either by the flap of a butterfly's wing or through the perseverance of Henry Enfield Roscoe, whence physics gained its lasting toe-hold in Manchester. It was most certainly not by the contribution of the first man to teach physics at Owens, Professor Archibald Sandeman, who was such an abject failure that, just like his successor Bellamy Clifton, who mercifully fled to Oxford, the mention of either names in Manchester or Oxford is met with ignorance or embarrassment. Let us not leap ahead. Owens College, the *Manchester Guardian*, Henry Roscoe, Archibald Sandeman and Bellamy Clifton are all treats in store for the next chapter. Back in the middle of the 19th century, whilst the usual suspects were discussing, arguing or falling out terminally with friends, usually because of religion (either they conformed or they didn't), James Joule was in the process of changing physics as it was understood. Before Joule, however, let us remember James Nasmyth.

1836–: James Nasmyth

Setting up business in Manchester

James Nasmyth set up business in Manchester in 1836 at the age of 28. His father was Alexander Nasmyth, an illustrious portrait and landscape artist, whose portrait of Robert Burns is well known. Whereas Nasmyth the elder was an artist with a great interest in engineering, his son James made his career as an engineer and inventor, but could equally well have developed his own skills as an artist, photographer and part-time astrophysicist to seek any number of alternative careers. He tried to fit all of them in.

His first business premises were a former cotton mill in Dale Street to the North of Piccadilly, but when one of his massive machine tools burst its way through the floor into the premises of a glassmaker below, it was time to set up on a six acre site in Patricroft, on the West side of the city owned by 'Squire Trafford', near what is now Trafford Park. The huge factory that he built from bricks fired from clay dug from the site endured, was taken over the Royal Ordnance in 1940 and finally demolished in 2009.

The squire offered terms of $1\frac{3}{4}$d annual rent per sq. yard, about £200 if Nasmyth's initial estimate of 6 acres was correct. Finance had been assured during his earlier reconnaissance of Manchester, when he met two of the Grant brothers[11], William and Daniel over dinner, which of course took place at 1 pm, it being Manchester. After assuring Nasmyth he had £500 credit at 3%, whenever he needed it, William Grant gave what Nasmyth described as an unforgettable knowing wink. Only two years later, now set up in Manchester, did Nasmyth discover that William had a glass eye which needed occasional facial contortions to keep it in his socket.

Nasmyth's first business premises in Manchester can be seen in Figure 1.27 on the next page and are shown for a manifold of reasons. Nasmyth drew the scene himself and it is useful to calibrate his accuracy as a combined artist and draughtsman and also because it gives a view of Manchester during the first half of the 19th century. It is located just to the North of Piccadilly in the now redeveloped 'Northern Quarter' near Stevenson's Square, which is now the centre of a conservation area. The

---

[11]The Grant brothers were immortalised by Charles Dickens as the 'Brothers Cheeryble' in Nicholas Nickleby.

map serves to orientate the viewer. Dale Street runs across the front of the picture and Newton Street runs diagonally off to the left in front of the main building shown, which was owned by the firm of Wren and Bennett, from whom Nasmyth rented a flat, which was the name then for a floor of a warehouse or factory. Wren and Bennett's main premises can be seen on the left, next to the 'Dale Street' street sign. Both of these premises are marked in pale blue on the map. Using Priestley's book on perspective [59] the location of the artist can be reconstructed and it is marked by the red symbol $\otimes$ on the map.

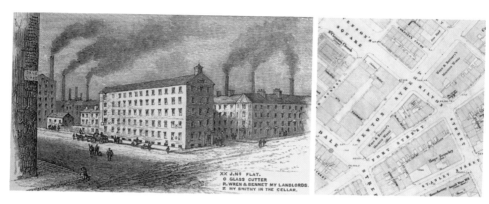

Figure 1.27: *Left: Nasmyth's first business premises on the corner of Dale Street and Newton Street. Right: A reconstruction of the area from contemporary maps.*

Because the coordination between what Nasmyth saw and what he drew was called into question regarding sunspots, a comparison of his drawing of the warehouse and the map is instructive. As drawn, Dale Street continues to the right in a straight line past the end of his building, past the entrance to Port Street and along the front of the block on the right (domestic houses), whereas the map clearly shows that Dale Street bends significantly at Part Street. By straightening out Dale Street in his drawing, Nasmyth was able to sustain, for artistic reasons, the eye-leading angle that runs from the bottom of the left hand corner to where it intercepts the right hand edge of the image, exactly one golden artistic third of the way up the edge. The map shows a lamp-post (L.P.) at the right hand edge of the end of his building and Nasmyth drew it. However, the lamp-post on the map shown on the corner of the left hand building by the street sign, is omitted by Nasmyth because it would have been artistically intrusive. The extremities of Nasmyth's flat are marked by × on the drawing. The conclusion that must inevitably be drawn is that we shall need to accept that Nasmyth employed licence in his art and the question, which it might not be possible to resolve, is that to accept his science, we shall have to accept that he suspended his licence when drawing sunspots and Mars. It can be proved that he certainly did when drawing the moon, because his 'photographs' of the moon are commensurately immaculate.

As an indication of what Manchester was like in 1840, Stevenson's Square (top left hand corner of the map) and buildings around it, had been established in the latter half of the 18th century to provide relatively affluent housing for the increasing number of affluent businessmen. Most of the streets in the area are named after entrepreneurs, such as William Stevenson himself, Ashton Lever, Dale, Newton, Anthony Bradley, not to mention Adam Oldham, after whom Oldham Street is named, despite it pointing in the general direction of Oldham. By 1840, the area was awash with mill chimneys and factories.

Nasmyth's first premises were pulled down not long after he vacated them and the subsequent replacement warehouse on the former site was built in a complete wedge shape to fill the available land. The new building, erstwhile called 'Bradley House', but later nick-named the 'Flat Iron', stands today as a listed building, now converted to a (budget) hotel.

Nasmyth's endeavours and achievements as a machine tool magnate were spectacular and his own painting of his massive steam hammer earns as much awe as the steam hammer itself. He set up house in Patricroft, six minutes walk from his factory and enjoyed passing through apple orchards and breathing the clean air as he commuted. The air especially would be conducive to his imminent career as an astrophysicist.

It is a matter of some note that Nasmyth, whilst starting and running a business that produced massive steam hammers, selling hundreds around the world as well as a similar number of railway locomotives, was able to do more physics and astronomy research and write more papers to the Manchester Lit. and Phil. than the first three physics professors at Owens, between 1851 and 1871. He ran his business from Manchester for 25 years and became sufficiently wealthy that he was able to retire and concentrate fully on the things that he liked doing such as natural science.

Observations in astronomy and astrophysics

Nasmyth's contributions to astrophysics came at a time when photography itself was in its infancy and the mere capture of a hazy image of the only feasible celestial object – the moon – was regarded as a triumph. To publish a paper of worth in astronomy needed a quality telescope, a keen eye, a brain whose imagination could be kept under control and good drawing skills. Nasmyth had all of these and in addition, he was able to cast and grind his own mirrors to the highest standards, having learnt how to do this in his bedroom as a child in Edinburgh. Once settled in his new house and business in Patricroft, he worked his way up through telescope inches until he reached a practical limit of 20 inches for the size of his garden, as can be seen in Figure 1.28, a device gazed at in puzzlement as bargees drifted along the Bridgewater Canal past his house.

Figure 1.28: *James Nasmyth's house in the future Trafford Park.*

He had a tendency to wear his night-shirt whilst observing and since he was always active during full moons, he was taken for a ghost by passers-by. He would carry smaller telescopes than his 20 inch monster around the garden and they were thought to be coffins.

Although dwarfed by William Parson's (3rd Earl of Rosse) 72 inch Leviathan of Parsonstown, it made up for its inferior diffraction limit by the most precisely ground mirror in Britain at the time. Regularly observing the lunar surface from Manchester, he soon realised that there was a serious deficiency in the published images of the moon. Photography was incapable of the detail and Nasmyth was not impressed by the quality of lunar cartography when undertaken by even the best astronomer, as he laconically remarked in his autobiography:

'Schröter was a fine observer, but an indifferent draughtsman.'

He came up with an impressive solution, which is described in detail in his book on the moon [60], co-written with his friend and astronomy colleague James Carpenter who was at the Royal Observatory, Greenwich. Eventually, after years of assiduous observation and drawing, examining the minutest detail of each land-mark and seizing every favourable opportunity to educate the eye, he accumulated fine drawings of the whole surface, with the length of shadows and angle of the sun recorded. He then proceeded to prepare plaster models of all the craters, the mountains and finally a full scale plaster model of the moon itself was constructed and photographed at leisure as it was illuminated by the actual sun. One can easily be deceived, if the accompanying text is not read, that Nasmyth's 'photograph' of the full moon shown in the centre of Figure 1.29 is of the moon itself, so well does it compare to even the most modern photographs today. Nasmyth was far ahead of his contemporaries in imaging the moon.

Figure 1.29: *Left: James Nasmyth observing with his 20 inch telescope in his garden in the future Trafford Park, almost certainly a self portrait. Top centre: Nasmyth's 'photograph' of his plaster model of the moon. Bottom centre: A modern photograph of the moon. Top right: Nasmyth's drawing of Mars. Bottom right: A modern photograph of Mars, from space.*

Nasmyth's reputation as an industrialist, helping to drive the industrial revolution in Manchester and the producer of the most graphic images of the moon ever seen, spread and came to the attention of royalty. When Queen Victoria and her consort visited Manchester in 1851 (see more below) they were guests of the Earl of Ellesmere at Worsley Hall. Prince Albert had seen Nasmyth's images at the Great Exhibition a few months earlier and Nasmyth was summoned the short distance to the Hall to bring his drawings and explain them to his Highness. Not surprisingly, as Nasmyth narrated in his autobiography [62]:

> 'Her Majesty took a deep interest in the subject, ... The Prince Consort said that the drawings opened up quite a new question to him, which he had not before had the opportunity of considering. It was as much as I could do to answer the numerous keen and incisive questions which he put to me. They were all so distinct and cogent.'

The Queen was reciprocatingly fulsome, as she wrote in her diary:

> 'The evening was enlivened by the presence of Mr. Nasmyth, the inventor of the steam hammer.'

before dwelling on his charming manner, which combined simplicity and modesty with the enthusiasm of genius.

Mars was another obvious target for Nasmyth and seizing the opportunity of a favourable conjunction of Mars in September 1862, he made several colour drawings and paintings of what he saw, as well as describing them in a paper to the Lit. and Phil. [61]. He also sent copies of his images to the geologist John Philips, who presented a paper to the Royal Society, published in 1865. The Royal Society archives contain a watercolour, some 4 inches square and their archive notes state it to have accompanied a letter to Philips dated the 20th of January 1863. The Natural History Museum in Oxford possesses a 'gouache' painting of Mars 77×56 cm which is described as being from Nasmyth and 'almost certainly used by Philips to accompany his paper to the Royal Society'. Philips did not mention Nasmyth in his paper, but referred to additional material placed in the Royal Society archives. There is no knowledge available whether or not Philips employed a local artist to magnify Nasmyth's small icon and in any case, unlike NASA who place all their images free from copyright into the public domain, these two images are not. Fortunately, Nasmyth's original was reproduced in colour in his Lit. and Phil. paper using the new and revolutionary 'Woodburytype' process and after more than one and a half centuries, this is now firmly in the public domain – see Figure 1.29 on the facing page. Close examination of the image shows it to be probably a chalk or pastel drawing. Woodburytype images were notable by their high quality and were rendered obsolete purely on the basis of cost when the cheap half-tone process took over. A special feature was the permanence of the coloured inks and with the Lit. and Phil. reproduction of Mars having spent most of its time since 1865 enclosed within a volume, the colours are likely to be close to those printed. Nasmyth described the large red areas but was also captivated by areas of blueish green which he assigned to seas. Given that Mars has large areas of red and current images made with good colour balance show other areas of neutral grey, it is possible that Nasmyth's eyes and cerebral colour processing cell, looking at Mars and Mars alone, fell victim to the established phenomenon of 'colour constancy' and the perception of areas of red and grey were perceived as red and green-blue. It was no fault of the observer and being male, there was also the high probability of Nasmyth being red-green colour blind, like I am.

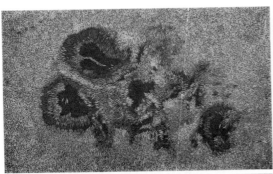
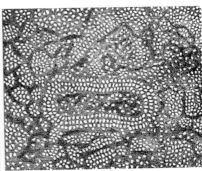
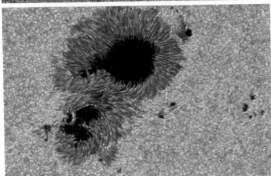

Figure 1.30:

Top Left: Nasmyth's 1862 drawing of a sunspot, its filamentary penumbra and granular surroundings. Top Right: William Huggins 1877 version. Bottom Left: A modern NASA image of the same.

The weak link in the production chain for many astronomers at the time was the infancy and inadequacy of photography. The human ingredient could introduce flaws as the episode regarding canals on Mars showed at the end of the 19th century. Nasmyth had exceptional drawing and painting skills, to which he added meticulous precision. He nevertheless became embroiled in a controversy regarding sunspots which still can arouse subjective judgements today. No one could deny that his detailed drawing of a cluster of sunspots, compared to a modern image made by NASA, shown side by side in Figure 1.30, contains all the structure that is still not fully understood today. He did not discover the penumbral

filaments which were observed by Johann Hieronymus Schröter in
and William Herschel in 1837 and possibly others before them. But he saw
more and on a clear day in the future Trafford Park, with low atmospheric
turbulence, he exulted to see a granular structure on the surface of the sun,
outside the sunspot region.

The ensuing row, precipitated by the astronomer and clergyman
Rev William Rutter Dawes, pulled Herschel and Warren De la Rue in
support of Nasmyth on the one side and Huggins and Lockyer aligned
with Dawes on the other. Dawes had acquired the nickname 'Eagle Eye'
but now it was largely a matter of language. Nasmyth had started off by
describing the granules as 'willow-leaf' in appearance and later conceded
that 'rice-like' might have been better. Dawes was not having any structure
at all because he couldn't see any through his telescope. Time has shown,
as the NASA images clearly show and what advanced photography already
showed early in the 20th century, that there is a clear granular structure
which tells a lot about the dynamics and geology of the sun. Hindsight
is the food of historic reviews and there were faults all round. There is
now no way to decide whether Nasmyth's eyes were deceived, possibly
by astigmatism or whether his imagination felt compelled to link the
penumbral filaments which clearly were elongated, to the granules, only
some of which were. One can accept the term 'rice-like' provided it is
agreed which kind of rice they resemble. Long grain Patna they are not.

It is a matter of opinion, but in mine, a 1997 review by
C F Bartholomew of the discovery of granulation [63] does Nasmyth a
disservice and gives Dawes more credit than he deserves. Nasmyth is
contemptuously dismissed as a 'retired engineer' whilst no judgement is
made on the prime occupation of the Reverend Dawes whom Bartholomew
presents as 'an experienced and respected solar observer'. Whilst pointing
out that Dawes launched his opposition to Nasmyth on the basis of
second-hand information, not having read any of the papers, Bartholomew
still did not credit Nasmyth with the discovery of granularity, which is
manifestly at odds with Figure 1.30. Nasmyth estimated their size at about
1000 miles compared to the modern accepted figure of 1500 km. What
more would anyone want? Despite the occasional bias that bubbles to the
surface, Bartholomew's paper is a good record of the historical facts that
evolved at the time. Dawes was adamant that apart from sunspots, the sun's
surface had uniform luminosity. Nasmyth was equally adamant it did not.

Bartholomew inexplicably assigned the discovery of granularity to
William Huggins on the basis of the image shown top right in Figure 1.30
on the facing page. Had Nobel prizes been around in 1870, Huggins would

not even have shared it. For all of his contributions to astronomy, Nasmyth had the honour of an eponymous crater on the moon. William Rutter Dawes had a crater on the Moon named after him as well as one on Mars.

At the age of 48, Nasmyth decided that he had made enough money to retire on and re-located to Kent to pursue his hobbies of astronomy, astrophysics, writing and painting. One of his final ideas was to devise procedures for turning cast iron into steel and it eventually led to the Bessemer process. Recognising his contribution, Henry Bessemer offered a third share in the patent, but Nasmyth decided he had made enough money and gave up engineering as an occupation.

1837–: James Prescott Joule

Although John Dalton and James Joule brought Manchester to the centre of the world stage of science during the first three quarters of the 19th century, their productive research careers did not overlap. In December 1834, only a few days before his 16th birthday on Christmas Eve, Joule started attending classes held by Dalton in a room provided by the Lit. and Phil.

Figure 1.31:

John Davies in 1833. Derived from a monochrome mezzotint engraved by S W Reynolds Sr, from a painting by S W Reynolds Jr, printed by the Manchester engravers Agnew Zanetti. Fifty monochrome prints were put into the future public domain.

Colour processing © 2017 Robin Marshall.

Initially he was taught arithmetic and geometry, with the first book of Euclid as the standard textbook and he attended for two classes of one hour each per week. Joule continued with these classes, which extended into the whole of the natural sciences, until 1837 when Dalton suffered a stroke. Thereafter Joule was taught privately by John Davies (see Figure 1.31 on the preceding page), a lecturer in mathematics, chemistry and natural philosophy at one of the two medical schools, specifically the one that had been set up in 1824 in Pine Street near the infirmary in Piccadilly.

Davies was usually present when committees were set up to form new institutions or colleges (see above) and was also a vice-president and lecturer at the Mechanics' Institution. John Davies was of sufficient precocious reputation or means to have a portrait of himself painted by Samuel William Reynolds Jr in 1840 and Figure 1.31 on the facing page is derived from a mezzotint made by Samuel William Reynolds Sr. By its content, the portrait demonstrates the importance of experimentation to Davies, whence the 19 year old Joule also made it the core of his career.

Joule's first paper

On the 8th of January 1838, two weeks and a day after his 19th birthday, and now taking classes with Davies, Joule sent off his first paper to William Sturgeon who published it in volume ii of his *Annals of Electricity*. The paper described an electro-magnetic engine he had built, and he imagined that he had 'succeeded in effecting considerable improvement in the construction of the magnets.' Cardwell [15] said that it could barely overcome friction. Cardwell went on to fill 333 pages with his superb biography of Joule and only highlights need be offered here.

Indeed, Cardwell's book is a model for anyone writing a biography. Joule's deficiencies and mistakes are described with candour and with the same emphasis as his successes. An example of the other sort of biography is Maurice Crosland's eulogy on Gay-Lussac [64]. Crosland appears to have become so dazzled by Gay-Lussac that his subject could do no wrong. The passage describing the relative priorities of Dalton and Gay-Lussac is more biased than Macaulay's account of the siege of Derry, if that be possible. In the brief account of the Dalton-Gay-Lussac law that I gave above, it was easy to ignore Crosland, because his science was wrong.

After a few further 'warm-up' and forgettable papers on electricity published in Sturgeon's *Annals*, the young Joule sent a paper to the Royal Society describing a new law of physics – that the heat W, generated by passing an electric current I through a wire of resistance R was given by equation $W = I^2 R$. The paper was rejected for publication in the Society's *Transactions*, an act akin to a music publisher rejecting young Beethoven's 1st Symphony. At the age of 21, Joule had produced one of the fundamental, true and lasting equations in electricity, conceived and born in Manchester.

Figure 1.32:

James Prescott Joule at the age of 45. The original oil painting was destroyed during the Manchester Blitz, Christmas 1940 but has been repainted by the author, using the engraving published in Osborne Reynold's obituary. This process used the author's inverse matrix techniques, Tikhanov regularisation, the Kuhn-Tucker algorithm, portraits by the same painter and the natural colours of the actual apparatus, now held by the Kensington Science Museum.

Colour processing © 2017 Robin Marshall.

Joule now began to present his work before the Manchester Lit. and Phil., impressing Dalton, and began his investigation into the mechanical equivalent of heat, for which he is best known. His first experiments were already ingenious. Water was heated by what was essentially an electric immersion heater with a motor included in the circuit. By using a system of pulleys and weights in pans to rotate the motor at the same speed as that produced by electricity he calibrated the power of the motor in units of mechanical force multiplied by distance. These and the subsequent experiments with weight and pulley driven paddle wheels enabled Joule, in the few years following his 24th birthday to establish, not only that the mechanical equivalent of heat was a universal constant of physics, but also to fix its value accurately.

The mechanical equivalent of heat

The debate on priority that followed Joule's papers and the one by Julius Robert Mayer [65] as to who first gave the world the mechanical equivalent of heat and the concept of energy conservation, overlooked the fact that only Joule actually measured it and showed it to be a universal constant.

Joule snorted when he read Mayer's paper, as I did when reading it for the first time. Prussian and German nationalism was on the rise and Mayer found support from the eminent chemist Justus von Liebig and the equally eminent physicist Herman Helmholtz. Liebig had no choice, having published Mayer's paper, despite it not meeting normal publishing standards, in his *Annalen der Chemie* and was thus bound by it. More surprising was the stance taken by the German educated Irishman John Tyndall, who deserves praise for not allowing his opinions to be dictated by nationality. All of Ireland was part of the United Kingdom from 1801 till 1922. But Tyndall took it to excess and could well have been the prototype for whom Gilbert and Sullivan put on their little list 'the idiot who praises, with enthusiastic tone, all centuries but this, and every country but his own.' Tyndall, no idiot, met his match when Manchester's Henry Roscoe dissected both Mayer's paper and Tyndall's reviews of the subject before the highly critical readership of the Edinburgh Review [66], whereafter a principle was set in concrete, that any acceptance of a new law or understanding of physics required irrefutable experimental confirmation and not just philosophical ponderings. Engineers already knew that.

One engineer in particular, John Smeaton was responsible for the foundations of the understanding of the interchangeability and conservation of different kinds of energy. He laid this out in two papers published in 1759 and 1776, the first one 80 years before Joule and Mayer. He may not even have been the first, but simply by considering him alone, all arguments concerning Joule's and Mayer's relative priorities are rendered obsolete and unnecessary.

Smeaton was neither prepared to design nor build anything unless he himself had made all the necessary physical and engineering *measurements* that underpinned his design. He is more famous for the Eddystone Lighthouse and innumerable bridges and harbours, than he is for his (underrated) contribution to the concept of energy and its conservation [67], and would not have used a cement for the lighthouse that was based on philosophical conjecture nor, and as will be revealed shortly, the contemporary incomplete theory of Newtonian mechanics. He varied the parameters of his cement, one at a time, until he converged on one that would set under sea-water and then maintain its strength. He had such faith in his knowledge of engineering and physics, that he spent more than one night on the lighthouse during predicted storms. A portrait of him, including the lighthouse is shown in Figure 1.33.

Figure 1.33:

John Smeaton, ca 1788.

Colour processed and re-lighthoused by Robin Marshall © 2017 using a contemporary monochrome mezzotint and Smeaton's own engineering drawings. The original depiction of his lighthouse in the engraving was seriously inaccurate.

John Smeaton's link to Manchester is twofold. His two papers mentioned above were reviewed and promoted enthusiastically in 1829 [68], by the long-term (forty years) Manchester resident, Scottish born, engineer and physicist Peter Ewart (see Figure 1.34 on the facing page) who castigated philosophers and mathematicians alike for deserting the debate on momentum and energy. Even today it is commonly held that Newton said 'Let my mechanics and gravity be.' But just as his theory of gravitation needed tweaking by Einstein and will probably need further tweaking in order to unify with quantum mechanics, Newtonian mechanics, heavily founded on momentum, needed tweaking to accommodate the concept of energy. Although Newton's blinkered disciples sought to fight the tweaking, Manchester was up to the task.

Ewart even complained that he was too polite to say what he really thought and that his targets would probably let him know if he was wrong. He wasn't. Newtonian theorists were about to accept, as late as the middle of the 19th century, the almost unpalatable truth that mass multiplied by the square of velocity was as important as mass multiplied by velocity.

$$mgh = \tfrac{1}{2}mv^2$$

Figure 1.34:

Peter Ewart in 1843. Derived from a contemporary monochrome mezzotint.

Colour processing © 2017 Robin Marshall.

The second link of Smeaton's to Manchester is that his undershot water wheel model (see Figure 1.35 on the next page) contains all the vital ingredients of Joule's eventual Mancunian mechanical equivalent of heat apparatus: the paddles, the strings, the pulleys and the tank of water, and most striking of all, the calibration of the prime source of energy (flowing water) by a mechanical system of pan weights and pulleys.

Ewart found it necessary to remind the scientific world what Smeaton had said 70 years previously, based on meticulous measurement. He had defined 'power' to be the mass of water multiplied by the head of the water (i.e. potential energy), and 'effect' to be the mass multiplied by the square of the velocity (i.e. kinetic energy).

Smeaton then declared unambiguously the relationship between potential and kinetic energy at a time when blinkered Newtonianists were denying the relevance of what is now called energy and the less blinkered didn't know the difference between energy and momentum:

'It therefore directly follows, conformably to what has been deduced from the experiments, that the mechanic power that must of necessity be employed in giving different degrees of velocity to the same body, must be as the square of that velocity.'

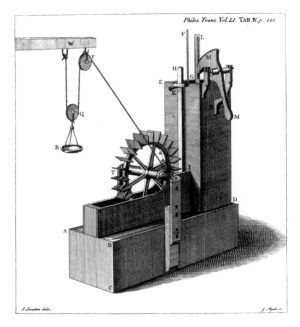

Figure 1.35:

John Smeaton's energy conservation paddle wheel apparatus, drawn by himself and engraved by the Royal Society's engraver, James Mynde.

In Table 1 of Smeaton's first paper, he converted the kinetic energy of a mass of moving water (he called it *effect* $= mv^2$) at various velocities, to potential energy (called *power* $= mh$) and the two columns of numbers are exactly in the proportion prescribed by the equation $mgh = \frac{1}{2}mv^2$, seven decades before Coriolis put the half in $\frac{1}{2}mv^2$. The paper contains a clear demonstration that Smeaton understood and was stating the equivalence and convertibility of potential and kinetic energy and was even presenting the numerical value of the conversion factor. He was less concerned with this than with telling the farming world that the amount of corn they could grind in one day was proportional to the square of the velocity of their stream of water. He left it to denying and bewildered Newtonian millers to wonder why their two streams each with half the velocity of their neighbour's single stream only ground half as much corn. The concepts were revolutionary in their day, but were easy enough to confirm by anyone who was prepared to admit that Newton, although right on momentum, didn't tell the whole story if you wanted to run a profitable business. The fact that Ewart had to scream them out again 70 years later is testimony to how revolutionary they were. Smeaton's knowledge and statement of the principle of conservation of energy is implicitly contained in his knowledge and statement that all the original energy (*power* = potential energy) had to be accounted for in the eventual effects (*effect* = kinetic energy), *viz*, gravitational potential energy equalled inertial potential energy:

> 'In comparing the effects produced by waterwheels, with the powers producing them; or, in other words, to know what part of the original power is necessarily lost in the application, we must previously know how much of the power is spent in overcoming the friction of the machinery; and the resistance of the air; also what is the real velocity of the water at the instant that it strikes the wheel; and the real quantity of water expended in a given time.'

This is as clear a statement on the conservation of energy in the language of the day, and given the definition of his language, Smeaton's words and the meticulous measurements that he reported, already declared the conservation of energy, and measured it, in the century preceding Joule and Mayer.

Smeaton was a consummate physicist as well as a prodigious engineer, defining his terms and knowing what they meant. Later generations might have renamed them, but he knew what they meant. There is also a curious parallel with the ancient geometry of Euclid. It is often forgotten that the laws of geometry arose from measurements by geometers and from those measurements, the thoughts of philosophers were stimulated to create their theorems of geometry. Pythagoras' theorem came about because it was found to be true by virtue of physical measurements on the surface of the earth. But in the subsequent centuries, the theorems came to be regarded more and more as philosophical conceptions, as their experimental foundations became lost in time. After Newton, the concepts of momentum, and the forces that made momentum change, became the new philosophical gospel and even meticulous 'kinetic geometers' like Smeaton could not alter the mathematical dogma. Ewart, as incensed as anyone struggling with the blinkered could be, yet too reserved to show it, quoted the Rev Reid [69], from 1748, then a relatively unknown minister in the Scottish Church, now better known as Thomas Reid the philosopher. Reid had both feet firmly in the camp of definable experimental science and had been equally incensed by the post Newtonian floundering debate on force, momentum and what eventually became known as kinetic energy:

> 'I say then, that it is impossible, mathematical reasoning or experiment, to prove that the Force on a Body is as its velocity,

without taking for granted the thing you would prove, or something else that is no more evident than the thing to be proved.

...

The controversy about the force of moving bodies, which long exercised the pens of many mathematicians, and for what I know is rather drop'd than ended, to the no small scandal of mathematics, which always boasted of a degree of evidence, inconsistent with debates that can be brought to no issue.'

It is difficult to read more than a sentence of Reid without warming to his mind and it comes as no surprise to find that he was the founder of the *Scottish School of Common Sense*. He became a professor at Aberdeen, four years after this supreme paper was written. His overlap with the mentality of practical physicists, Smeaton, Ewart and Joule is as great as his connexion to the mentality of Mayer is small.

Ewart himself constrained his exasperation within the bounds of polite language:

'In answer to the objection implied, in the reasoning of M. Laplace, against the force being as the square of the velocity, I can only repeat, what I have already so often repeated, that it is not the pressure exerted in a *given time*, but the pressure exerted through a *given space*, that is understood to be universally as the mass into the square of its velocity and I may add that there is nothing hypothetical in this conclusion – being derived from an induction of facts, it must stand or fall with the facts on which it is grounded.'

In this way, the convertibility of one form of energy (potential) into another (kinetic) and the numerical equivalence thereof, was already laid out by Smeaton and others, 80 years before Joule's experiments and Mayer's conjectural thoughts. It was left to Count Rumford, in two papers of beautiful content, but unnecessary length and turgidity [70], to declare the universal convertibility of kinetic energy into heat, observing and measuring the effort and resulting elevation in temperature when the bores of a number of cannons were drilled out. By measuring the heat capacity of the swarf, he cleverly and simultaneously demolished the caloric theory. Joule later showed that Rumford's numbers implied a value for the mechanical equivalent of heat that was of a similarly poor accuracy to that of Mayer's guess. The existence of such a universal constant did not occur to the manifestly diligent Rumford.

When all uncited thoughts, which can be traced to predecessors, are discounted in Mayer's paper, and there are many, there possibly remains the bright idea that the ratio of specific heats of a gas, which he had not measured, could be used to calculate the mechanical equivalent of heat, which he then did. The assumption, although needing verification by Joule many years later, already fell foul of the first half of the quotation of Reid above. It fell even more foul of Smeaton's declaration above, that if you start with some energy, you must account for all of it at the end. Mayer ignored all possible ways that the compressive energy could be dissipated except the one that produced heat. He could have read Ewart's paper which had been around for 10 years, but he didn't. Ewart, in his January 1829 paper [68], had already shown why Mayer was wrong, even anticipating Mayer's argument by over 10 years. The assumption that Mayer made to prove his point, could not be assumed at the time. It was at that time, unproved that *all* the work done in compressing a gas went into raising its temperature, thus appearing as heat. Ewart's model of a gas [68], (see Figure 1.36) a collection of interacting particles, was significantly ahead of its time. It is still used, even graphically, today.

Ewart's model has springs between the particles of gas, whereas Dalton's model of a gas assumed no inter-atomic interactions. These 'springs' are now known as Van de Waals forces, although their strength in air at room temperature is below the level measurable by Joule. Ewart used his model to explain why superheated steam cooled dramatically when forced at high pressure through a small nozzle. This early demonstration of the Joule-Thomson effect has also faded into obscurity. Joule and Thomson needed many experiments and much thinking to put the process on a sound basis to the extent that it could be used to liquefy hydrogen and helium.

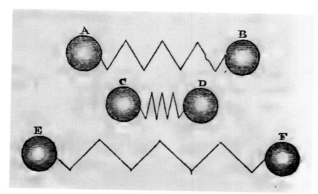

Figure 1.36:

Peter Ewart's 1829 model of gas particles.

Ewart argued that gas particles in equilibrium at atmospheric pressure would be like A–B in Figure 1.36. When the gas is compressed, the spring is compressed and becomes like C–D. Then when the gas escapes through a nozzle suddenly, the negative impulse causes the spring to expand to E–F, beyond its original equilibrium position and energy must go into the expanding spring, causing the gas to cool. This is, in fact, the same argument used by Mayer 10 years later to equate the work done in compression to the increase in heat.

Mayer's paper was now essentially stripped of all originality. Roscoe was right; Tyndall was not. Of course, Ewart's argument was overcomplicated. If the gas at C–D is allowed to reach equilibrium, a sudden expansion to E–F will cause cooling irrespective of the original atmospheric state A–B. The cooling is determined by the adiabatic gas law which contains the ratio of the specific heats. Mayer's idea was a good one philosophically, but as Smeaton had showed, philosophy does not always grind corn profitably.

Roscoe's annihilation of Mayer and Tyndall in his Edinburgh Review also cited Humphrey Davy's experiment of 1799 where two pieces of ice, in a room below 32 F, melted when rubbed together. Mayer had included this phenomenon in his philosophical argument as well as Davy's demonstration [71] that wax could be melted by rubbing two metal plates together, as if he had done both experiments himself, even using the word 'We' when describing what Davy had done. Indeed, Davy's paper in 1799 [71] contained almost as much philosophical conjecture on the equivalence of frictional mechanical motion on a body and the heat gained as did Mayer's.

Mayer's paper contained one remaining feature, which, like all other content was presented as an original conception. As ship's doctor on an exotic voyage to Indonesia, Mayer noticed the difference between veinal and arterial blood. There is nothing in his observations, reported in 1842, that cannot be found in the 2nd edition of Adair Crawford's book *On animal heat* [72] published in 1788. Before we place a halo on the head of Dalton's hero Crawford, let us not forget that, albeit in accordance with the *mores* of the times, he also published a paper on the effect of port wine on stomach lining muscle by carrying out unspeakable experiments on kittens, in the days before guinea pigs and sacrificial rats. The journal's publisher subsequently apologised.

Other distinguished scientists had also contemplated the conservation of energy long before Joule and Mayer. Thus, Ainé Seguin wrote to the *Edinburgh Philosophical Journal* in 1824 [73] giving and explaining a principle that his grand-uncle, Joseph Montgolfier had often discussed with him:

> 'The principle which he maintained was, that the *vis viva* could neither be created nor annihilated, and consequently, that the quantity of motion on the earth had a real and finite existence.'

Seguin's *vis viva* was essentially kinetic energy, but the discussion included the interesting conjecture that (upper atmospheric) gases were cold because they had lost kinetic energy overcoming gravity. It is therefore clear that conservation of energy and even its equivalence to heat had been widely discussed long before Mayer mentioned it again in passing. Seguin also wrote a book on the energy and finances of railway engines [74], which Roscoe used in his demolition of Tyndall and Mayer.

Joule and Seguin were on good terms, if the opening paragraph of a letter to Joule [75] be regarded as a friendly jibe. Joule felt able to present the actual letter at a meeting of the Lit. and Phil. on the 12th of December 1862 and to have it printed in the proceedings:

> 'Dear Sir,
> I am much obliged to you for the disinterestedness you have maintained in reference to the discovery of the great principle that my uncle, the celebrated Montgolfier, revealed to me some sixty years ago. Since that time I have passed my whole life in studying this question.'

By 1845, Joule had used four quite different methods to determine the mechanical equivalent of heat and in his paper of that year, presented to the British Association meeting in Cambridge, he extended his new theory of heat to gases and extrapolated to zero pressure in the way shown above for Dalton and Gay-Lussac's data in Figure 1.14 on page 51. Joule obtained exactly the same value of -267 C.

At the age of 27, Joule's scientific achievements already included the electrical ohmic heating law, the establishment and measurement of the mechanical equivalent of heat, major contributions to the law of conservation of energy and the first publication of the value of the absolute 'zero' of temperature. The remainder of his scientific career can be followed in Cardwell's book. I now present some other aspects of his life, not widely reported elsewhere.

Joule's entry into adulthood and scientific research coincided with the invention of photography, in which he became interested. Rather than presenting one of his pictures here, I show a photographic portrait of him in Figure 1.37 on the next page taken by Henry Roscoe's wife, who was an accomplished photographer herself. It is printed in Roscoe's autobiography [8]. The photograph has been colour processed using matrix transformation algorithms which I developed and published in the Journal of the Perthshire Society of Natural Science [17].

Joule was a natural choice as Professor of Natural Philosophy at the newly founded Owens College in 1851, but as we shall see in Chapter 2, he turned down every opportunity to pursue an academic career although he thoroughly supported the college when it strove to become a university and joined the delegation which travelled to London in a vain attempt to try and persuade prime minister Gladstone to relent (see also Chapter 2).

Figure 1.37:

Joule in later life, image derived from a photograph by Mrs Roscoe.

Colour processing © 2017 Robin Marshall.

Statues

Soon after Joule's death on the 11th of October 1889, the Manchester city authorities agreed to accommodate a permanent memorial to their man of heat in the form of a white marble statue which stands inside the doors of the Town Hall. Joule is on the right as one enters the building, directly opposite the statue of Manchester's other man of great stature, Dalton. Joule's statue was sculpted by Alfred Gilbert, a member of the Royal Academy. Of the millions who have passed Joule's statue since it was put in position over 100 years ago, the majority will have walked past without looking, knowing or caring who he was. A few will have glanced sideways and wondered who this could be, depicted in white marble and holding small scientific instrument in his hand. Some will have walked round the statue and noticed the name 'Alfred Gilbert 1893' on the back and then out of curiosity walked round Dalton's and found the name 'John Dalton' on the back of that one. The inconsistency will provoke the very few to think that Dalton was the name of the man who made the sculpture (rather than Chantrey) or that Joule's statue is of Gilbert. A small number will know the identity of both statues and sculptures.

Figure 1.38:

The marble statue of James Joule by Alfred Gilbert, in the entrance to the Manchester Town Hall.

Photo: Robin Marshall.

1840–1853: The Manchester New College (Again)

The Manchester Academy, once the Warrington Academy, which had re-located to York in 1803, returned to Manchester in 1840 on the retirement of Dr Wellbeloved and renamed itself 'Manchester New College'. The Academy did not re-occupy the building it had vacated 37 years previously, even though the premises still bore the title 'College Buildings' (see Figure 1.21 on page 60). Instead they located themselves in the oldest and grandest house in Grosvenor Square, All Saints, situated on the corner of Ormond Street and Cavendish Street.

The Manchester New College in this new incarnation is highly significant in its role of providing teaching in physics in Manchester. It offered local inhabitants the first opportunity to obtain a degree in science, including natural philosophy. A Royal Warrant, dated March 1840, gave

93

the Manchester New College the right to send its students for degree examinations at the University of London, putting it on an equal footing with University, King's and other colleges incorporated into the University. This right brought with it a responsibility to provide the necessary quality of teaching.

The syllabus of introductory lectures was published in book form [76] and included a course (discourses) by Montagu Lyon Phillips on Physical Sciences and Natural History. The book gives a brief history of the College, from the time of its establishment in 1786 with the intention to:

> '... afford a full and systematic course of Academic Education for Divines, and preparatory instruction for other learned professions, as well as for civil and commercial life; and which should be open to young men of every religious denomination, from which no test or confession of faith should be required.'

The Preface to the syllabus refers to the original founding of the academy in Manchester, with academical buildings being erected and courses delivered until 1803, when 'circumstances connected with the Theological department' occasioned the removal of the institution to York. The re-establishment of the college offered the chance to make changes to the organisation, which experience had shown desirable. The most important of these was the connexion with the University of London. By a warrant dated the 28th of February 1840, Her Majesty empowered the officers of the College to issue certificates to those who should have completed the requisite course of instruction, enabling them to become candidates for the degrees of Bachelor or Master of Arts, Laws and Medicine, conferred by the University of London. Students of physics gained an Arts degree. The best students were rewarded: 'The candidate who shall most distinguish himself in Mathematics and Natural Philosophy, at the Matriculation, shall receive from the University an exhibition of £30', thus facilitating the furtherance of gaining a degree. The College also sent out a message to prospective employers:

> 'It may reasonably be anticipated that those public bodies which at present confine their privileges and facilities to graduates of Oxford and Cambridge, will at no distant time grant them equally to those of the University of London.'

Care was taken to separate the theological department from the literary and scientific and any religious exercises and instruction of those not preparing for the ministry was to rest entirely with the students' friends.

A public appeal was made for subscriptions and benefactions, especially to help with the cost of the removal to Manchester, the hiring of more Professors and for philosophical and chemical apparatus far exceeding in extent and costliness what the College previously possessed. The College pointed out that Manchester was now as accessible from distant parts of the kingdom, as it used to be from remoter parts of the district in which it stood.

Montagu Lyon Phillips remains a relatively obscure Victorian physics professor; he died at the young age of 47, there being a gravestone of a person with that characteristic name, born 1814 and buried in St Andrew Parish Church, Hove. The Literary Gazette of March 1839 [77] contains a list of new books, including 'Lectures on Natural Philosophy, by M. L. Phillips, foolscap 5 shillings.'. He is also listed by Stephens and Roderick [78] as having given a course of eight lectures on Pneumatics at the Liverpool Mechanics' Institute in 1839. He was elected as a Fellow of The Chemical Society of London (formed in 1841) at their meeting on the 17th of March 1851. Phillips also came to the notice of the New York Times on the 19th of March 1855 in an article announcing his new book [79]:

> 'The *vexato questio* between Dr Whewell and Sir David Brewster will shortly receive another exponent, in the person of Mr Montagu Lyon Phillips, whose work on *Worlds beyond the Earth* is announced by Mr Bentley.'

Astronomers Sir David Brewster and William Whewell had engaged in public debate about the existence or otherwise of extra terrestrial life and Phillips joined in (see also Miller [80] and Faber [81]). In his introductory lecture of 1840 [76], Phillips mentions the honour he felt to be appointed Professor in the Department of Physical Science and Natural History at the New College. It was devolved to him to deliver Lectures on Mechanics, Pneumatics, Acoustics, Heat, Electricity, Chemistry, Hydrostatics and Hydraulics, Optics, the Classification of Animals and Vegetables, Animal and Vegetable Physiology, Structural Botany, Natural History and Geology. He explained the difference between theoretical and experimental physical science, preferring the French word 'experience' to 'experiment'. He placed at the head of Natural Philosophy, the subject of *Mechanics*, without which one could not construct or contrive our steam engines, lathes, cotton mills, ships, houses and an infinite variety of machines. After mechanics, he placed pneumatics (the physics of gases and vapours) next in importance, giving the steam engine as an example of a pneumatic machine. Montagu Lyon Phillips appears to be the first Professor of Physics in Manchester to teach the subject to University degree level.

Despite the manifest effort that Phillips devoted to the College, he is mentioned only once by Davis [32] in his 1932 history of the meanderings of the College and its staff, despite swathes of paragraphs and pages devoted to manifestly inept theology teachers. As the official historian to the College and devoted to it, Davis comes across as a rather biassed Macauley-like recorder of events and people, oblivious to anything that did not accord with his blinkered experience or prejudice. He was only interested in religion and the arts that supported it, unaware that for centuries, science, especially astronomy, had been included in the education of theologians, so that they could better understand 'God's creation'.

From 1840 to 1853, the New College in Manchester simply did not attract enough lay students to be financially viable (Mercer [82]). Warrington educated 338 students between 1756 and 1782, an average of 13 per year. Manchester had 121 between 1786 and 1803 at an average of 7.1 per year. In the 27 years from 1803, York attracted 113, just over 4 per year but during the 14 years of its second stay in Manchester up to 1853, there were only 28 lay students, an intake of just 2 per year. By 1847, there was already much anxiety among the governors and their attention turned to Owens College which was about to be created.

For three years the governors debated whether they should attach themselves to Owens and they set up a committee to consider the respective plans of University College in London and Owens. The committee met on the 17th of December 1851 to consider a report on the options. They noted that Owens' bequest stipulated religious freedom and although religious instruction was to be provided, they saw no reason to excite the apprehensions that were first felt. Eventually, Owens decided to set up three religious courses, albeit after much heated debate about whether there should be any at all. There was a course on the Greek New Testament, one on the Hebrew of the Old Testament and a third on *The influence of religion in relation to the life of the scholar*. Before the year was out, the New College governors had become sufficiently alarmed by these courses and they resolved to affiliate themselves with University College, London who offered no such instruction. There were internal disagreements on policy and the matter came before the Master of the Rolls (Sir John Romilly) and the Solicitor General (Sir Richard Bethell) was moved to argue on behalf of some of the trustees against union with Owens:

'But if there is to be religious instruction open to the young men provided for them in Owens College, there is no security that that would be according to the particular tenets of our body of Dissenters (Unitarian): and then, what are we to think of young men being sent to Manchester New College for the purpose of being educated in a certain set of religious principles, and then left at liberty to go to another institution where they would fall in with persons pursuing a different mode of religious instruction ...'

The action has to be judged by the religious *mores* of the age even though when viewed 160 years on, no one appears to emerge from the situation with credit. However, the biggest objection to union with Owens was stated by the committee to be the untested nature of the College, a view which can hardly be challenged and which would have appeared to be increasingly justified as Owens floundered in its early years. Hence in 1852, Manchester New College resolved to remove itself to London and did so in 1853, where it became located in University Hall, Gordon Square. It left the teaching of secular courses to UCL and became purely a theological college. Then in 1893 it moved to Oxford, opening its new buildings designed by the Unitarian architect Thomas Worthington in 1893. Harris Manchester College was granted Permanent Private Hall status in 1990 and became a full college of Oxford University in 1996.

1841–: John Benjamin Dancer

No account of 19th century physics in Manchester would be complete without mention of John Benjamin Dancer, who combined an ability for invention with the technical skills to pursue and develop whatever he invented. More than an instrument maker, he was at the same time, not only an expert in physical optics but also other branches of the science.

Brief genealogy

John Benjamin Dancer was born in London on the 8th of October 1812. His father was Josiah Dancer, an optician and manufacturer of optical, philosophical and nautical instruments. Here, the word 'philosophical' has the sense 'physics'; the word 'physics' not being adopted until 1834. J B Dancer's mother is of no mean interest. She was born Anna Maria Tolkien in 1779, the daughter of John Benjamin Tolkien, from whom Dancer inherited both his given names. Anna Maria's eldest brother was named George William Tolkien, who married Eliza Lydia Murrell. Among

George and Eliza's many children was their second son who was also called John Benjamin Tolkien and skipping a generation, John Benjamin's grandson was John Ronald Reuel Tolkien. It is therefore likely that John Benjamin Dancer's inventive genes derived from the Tolkien side. The illustrious J R R Tolkien himself will appear once more in a Manchester physics context in a later chapter. Both he and Dancer held the familial name 'John'.

Business in Manchester

Josiah Dancer removed his business from London to Liverpool in 1817 when J B was just four years old and upon the death of Josiah in 1835, John Benjamin took over the business at the age of 22. In Liverpool he found himself in competition with the more experienced, less technically competent, but better businessman, Abraham Abraham. The young John Benjamin was fired by the invention of photography and having read up on the details, produced a few daguerreotypes, spent three days traveling to London to a daguerreotype exhibition in 1839 and decided that his were as good as any. Dancer's microscopes were also superior to Abraham's and they solved their simultaneous presence in Liverpool by the following advertisement in the Manchester newspapers:

OPTICAL MATHEMATICAL AND PHILOSOPHICAL
INSTRUMENT   ESTABLISHMENT
13 CROSS-STREET, King Street, Manchester

– – – –

A ABRAHAM, of 20 Lord-street, Liverpool, respectfully announces in conjunction with Mr J DANCER

– – – –

they propose OPENING the above PREMISES on Monday the 21st instant. Their stock will embrace an extensive assortment of Optical Mathematical, Surveying and Mining Instruments ... a large collection of Pictures from the Daguerreotype;
13 Cross-street, King-street, Manchester, June 4th, 1841

Whether by prior agreement or the natural evolution of their business or even personal relationships, Abraham terminated his partnership after four years and continued his business thereafter from Liverpool, leaving the Manchester turf to Dancer. Both the partners made microscopes and Dancer sought to improve the quality of the instruments he marketed, and was successful in selling one to John Dalton during the years that elapsed between the establishment of his business in Manchester in 1841 and Dalton's death in 1844. Dancer also made thermometers and had an insatiable local customer in James Joule who gave them the accolade:

> ' ... the first which were made in England with any pretensions to accuracy.'

Dancer's lifetime work achievements

Once in Manchester, Dancer became an active member of the Lit. and Phil., publishing regularly at the same time as running a business. Many scientifically minded businessmen were able to do this, putting to shame the early the professors of natural philosophy at Owens College who published little.

Immediately on setting up his trade in Manchester in 1841, he started to sell daguerreotype apparatus and taught the process to John Dale, a chemist, and Joseph Sidebotham, a calico printer and dyer. According to Dancer himself [83]:

> 'Many of the Manchester gentlemen became amateur photographers, and it soon became a popular amusement. In November, 1841, Mr Beard, who had purchased the patent right in England, opened a daguerreotype portrait gallery in rooms over the Manchester Exchange. The late Mr Peter Clare, at my request, induced the late Dr. Dalton to sit for his portrait at this gallery. Three only were taken – Mr Peter Clare had one, Mr John Dale had another, and a third fell to my share; these were the only photographic portraits taken of the celebrated Dr Dalton.'

This brief quotation casts into doubt the provenance of several important daguerreotypes that are widely attributed to Dancer, but are probably not his. The first ever photograph taken in Manchester (see Figure 1.39) is widely assigned to Dancer. It was taken from the 'top' of the Manchester Exchange building, exactly where Richard Beard ran his daguerreotype business. Dancer sold many daguerreotypes from his own premises, a few strides away in Cross Street and he labelled them all 'J B Dancer' irrespective of who had taken the picture. The balance of probabilities is that Richard Beard took this photograph. Likewise, the daguerrotype reproduced on the right in Figure 1.39 on the next page, one of a stereo pair is also widely stated to be a self portrait of Dancer himself. There are several similar pictures of the same setting, with the same instruments, the globe and bucket and they are said to be of 'Scientist in his laboratory' by J B Dancer. Figure 1.39 may or may not be a portrait of him.

The provenance of the John Dalton portraits appears, at first sight, to be equally confused, although there should be none, given his description above of how the photographs came to be taken.

Figure 1.39: *Left: Daguerreotype of a Manchester street scene. Right: John Benjamin Dancer (possibly).*

On the 9th of January 1931, Henry Garnett, the son of the co-founder of the firm of Flatters and Garnett of 309 Oxford Road[12] penned a letter [84] to *Nature* which was published on the 7th of February. Garnett had published a brief article about Dancer in the *Memoirs of the Manchester Lit. and Phil.* [85] in 1928 and was now searching (vainly as it turned out because his letter elicited no reply) for the three photographs:

'I shall be glad if any readers of NATURE can assist me in tracing the present whereabouts (if still in existence) of three original photographs of Dr. John Dalton. These were taken in Manchester, at one sitting, somewhere about the year 1842, by the Daguerre process, then recently introduced into Great Britain, and so far as I know were the only photographs of the great chemist ever made. Their production has been wrongly attributed to John

---

[12]Flatters and Garnett sold optical instruments from premises at 309 Oxford Road, situated directly opposite the University buildings.

B Dancer, the fact being that it was through Dancer's good offices that Dalton was induced to sit at the local Daguerre studio. It is on record that one of the three copies passed to Dalton himself, another to Dancer, and a third to Mr John Dale, manufacturing chemist. Dancer's passed at the time of the Jubilee exhibition in Manchester in 1887 to Mr (afterwards Sir James) Dewar, the eminent chemist; I have also seen it stated that another (possibly Dale's) was in the possession of the late Mr Thomas Kay, manufacturing chemist, of Stockport. There is no trace of such a photograph in the collection of Dalton's apparatus at the house of the Literary and Philosophical Society where he did so much of his work, nor in the more personal relics preserved at Dalton Hall. Dancer's photograph was lent by him on various occasions to artists and engravers for copying, and became somewhat disfigured in consequence. Perhaps this letter may meet the eye of someone who has actually seen one of the originals or can assist me in tracing them.'

The copy retained by James Dewar appears to be now in the hands of the Science Museum in Kensington with digital copies available for purchase from Getty images and others. Curiously, and in clear contradiction to the easily available record in the public domain, they all attribute J B Dancer as the photographer. In a book [86] by A L Smyth, a librarian at the Lit. and Phil. in 1966, the daguerrotypes themselves are wrongly assigned to John Nicklin, about whom little else is known. Smyth also states that one of the original daguerrotypes was in the possession of the Society and another held by the Science Museum. The whereabouts of the third was not known. That the Lit. and Phil. held one of the three in 1966 is clearly at variance with Henry Garnett's letter. Yet Smyth asserted that a photographic enlargement was made and used in a paper [87] by L L Arden in 1961 and considerable reliance should be placed on this..

The Science Museum specimen is in rather a sorry state after having been lent to several engravers for copying. Dancer made it very clear above who took this photograph, which is reproduced in Figure 1.40 on the following page. The two most eminent engravers of the time were Charles Henry Jeens and James Stephenson. The cruder version by Jeens dates from 1842 and precedes the photograph. Stephenson's engraving (see Figure 1.40) was made in 1845 shortly after Dalton's death from drawings he made in 1842 and is clearly a direct copy of the photograph and a good one at that. Dalton's spectacles have been removed, which is surprising since they were a trademark of his appearance. Jeens version shows the mirror profile. A daguerreotype, by virtue of the optical inversion caused by the lens and then the rotation of the plate to make it upright, looks exactly as the subject when viewed from the camera. So Stephenson's version corresponds to the actual subject. Jeens flipped the subject for his own reasons. If any photo editing software is used to flip Jeens' image, place it and scale it to Stephenson's, it will be found that the features match precisely and only the hair and clothing differ. Stephenson took liberties with the shirt and jacket and Jeens took liberties with the hair.

Figure 1.40: *Left: Daguerreotype of John Dalton, made shortly after November 1841. Centre: An engraving by James Stephenson from the daguerreotype, 1845.   Right: An engraving by C H Jeens from the daguerreotype, 1842.*

J B Dancer's greatest claim to fame is his invention of micro-photography, which was provoked by a request from a friend who wanted a photographic reproduction of some family portraits, but much reduced in scale so that, for example, it would fit into a locket. Given the size of the portrait and the locket, Dancer would have needed but a few seconds with pencil and paper and the lens formula $\frac{1}{u} + \frac{1}{v} = \frac{1}{f}$ together with the magnification, $m = \frac{v}{u}$ to realise that he needed an achromatic lens with a focal length of about one and a half inches. He could have ground the lens himself because he possessed the necessary skills, but to save time and money, he bought one off the shelf; from a butcher's shop that is, as he explained himself:

> 'At the request of a medical friend I took a reduced picture of a family portrait, employing the eye of a recently killed ox for my camera lens.'

This launched a huge trade in miniature photographs and yet Dancer, failing to patent the idea as he failed to patent most of his inventions, did not become rich from something that was so far reaching. Many were those who subsequently laid claim to the invention of micro-photography, or had it made for them by over-enthusiastic supporters. Many were those who fought in the courts, claiming rights. Many were the laurel trees felled to adorn the heads of those who had no claim to fame, much as spare laurel was used to honour Werner Siemens for not inventing the dynamo and to praise Edison who did not invent the electric light bulb, with alas, non left over to laud him for inventing the electric chair. *'Palmam qui meruit ferat.'* declared George Shadbolt, as he advanced his own claims [88] to be recognised as the inventor, shamelessly even using the journal of which he was editor as a mouthpiece, whilst modestly also mentioning that the procedure was so obvious that it barely deserved mention. Also worthy of note was the photographer Renè Prudent Patrice Dagron who saw an exhibition of J B Dancer's microfilms in Paris in 1857. With a speed that Dancer never strove for nor was capable of, Dagron secured a patent in 1859. In 1870, he proposed that the beleaguered French army could get messages across German lines by sending microfilms by carrier pigeon. At 0.05 g each, a pigeon could carry 20 on a single journey. Not surprisingly, he is regarded as the Edison of micro-photography in France.

It is rare for inventors to receive awards and honours posthumously but in 1960, Dancer's great granddaughter received a posthumous Medal of Meritorious Service to the microfilm industry, from the National Microfilm Association of America. *'Denique palmam qui meruit ferat.'*

In 1852 Dancer invented a binocular stereoscopic camera based on an idea which had been brought forward by David Brewster in 1847. Up till then, a single camera was used to make the highly popular stereoscopic pairs of the time; the camera took a picture and was then moved sideways to take another. Some photographers inexplicably moved the camera by many feet, producing a bizarre optical effect. Dancer unsurprisingly showed that the optimum distance between the two lenses was equal to the average separation between two human eyes.

During the course of his electro-chemical researches, he discovered the basis of electrotyping by depositing copper electrolytically on an engraved copper plate. Then in 1838 he vastly improved the performance and durability of the Daniel cell, by replacing the ox bladders and gullets with porous ceramics. He also crimped the plates in his Daniel cells, thereby doubling its effective area. During his researches in this field, he produced ozone without recognising its significance.

He was also active in the overlapping areas of physics and electrical engineering. A neat invention of his was the subdivision of secondary

...dings of induction coils, thus allowing a choice of output voltages. He also devised the electric make and break circuit breaker. Its use in door bells alone would have earned him a fortune had he been alert enough to patent the idea. One of his inventions, the so-called fairy fountain, a multi-jet fountain illuminated from below with coloured lights and controlled by an electric keyboard, would have appealed to Adam Walker. A version was installed in the large fountain in the centre of Barcelona.

Although he is frequently credited with the invention of limelight, the original discovery and its development were carried out by others. What can be said with certainty is that he supplied the Manchester Mechanics' Institute with a limelight projection lantern of his own manufacture and was the first, or among the first, to produce photographic lantern slide transparencies which hitherto, had to be drawn by hand.

His achromatic microscopes, one of which was made for John Dalton, were renowned and were the main reason for establishing his reputation. Five years after parting from Dancer, rival microscope manufacturer Abraham Abraham suffered the pain of reading what the 1851 Crystal Palace Exhibition jury had to say about his: 'Abraham's microscopes on exhibit "are not such as demand special notice" '. He made improvements to telescope mountings, rain gauges, speed indicators, surveyor's levels and air pumps and invented an apparatus for Sir Joseph Whitworth for checking the accuracy of rifle barrels. As well as the precision thermometers he made for Joule, Dancer also made a traveling microscope so that the small differences in temperature could be measured with high accuracy. The Museum of Science and Industry in Manchester hold one of the traveling microscopes as part of their Joule collection.

Charity appeal

Around 1870, Dancer developed diabetes and his sight began to deteriorate, causing him to abandon his business. During his working life he was a regular correspondent with local newspapers, who printed his free advertisements in the form of letters to the editor. His granddaughter Eleanor Elizabeth, born in 1869, was eventually enlisted to write his letters and a lengthy one, more of an article *The introduction of photography into Liverpool and Manchester* was published in the Manchester newspapers and syndicated in the British Journal of Photography. Aware of his total blindness and current precarious financial status at the age of 73, a group of Manchester luminaries including James Joule, the aforementioned John Dale and two senior Manchester University professors W C Williamson

(Natural History) and Balfour Stewart (Physics) wrote a letter to the Editor of the Manchester Courier, setting up a charitable appeal to provide an annuity for Dancer. Referring to Dancer's own description of his contribution to the science of photography, the gentlemen listed his other achievements. They then went to the core of Dancer's lifelong self-inflicted problem and how they proposed to deal with it:

'It is sad to have to say, that notwithstanding Mr Dancer's talents and achievements, he is now living in very straitened circumstances, is moreover afflicted with almost total blindness, and therefore unable to follow the optical business to which his life has been devoted. It is not an unusual thing for a man of great mechanical ingenuity and skill to be an indifferent man of the world, and so it has been with him; as a business man he has been a failure. He has made improvement after improvement, invention after invention, any one of which might in "pushing" hands have made a fortune; but more interested in science than in money- making, he has allowed the golden chances to become public property, and has thus remained poor himself, while the world has reaped the advantage of his labours. Mr Dancer is now in his 74th year, and we beg respectfully to suggest that in this hour of darkness the world should pay back to him something for that which it has freely received at hands.'

The appeal subscription was launched although J B Dancer died not long after on the 24th of November 1887. With added poignancy to Manchester physics, Balfour Stewart, about whom much more in future chapters, died of a stroke three weeks later.

1843: The Lancashire Independent College

If one should walk along College Road in Whalley Range, Manchester today, and peer through the iron gates, one could hardly fail to be impressed by a magnificent edifice of Oxbridgian college proportions, set in spacious grounds. It is, or rather was, the Lancashire Independent College, designed to last a thousand years which it still might do, as a listed building. The painter Henry Edward Tidmarsh, more remembered for street scenes of London, also painted many local scenes to illustrate the 1890 book *Manchester Old and New* [3] by William Arthur Shaw. Figure 1.41 on the next page shows Tidmarsh's depiction of the College, shortly before 1890. It did not change in the next 127 years.

Figure 1.41: *H E Tidmarsh's painting of the Lancashire Independent College.*

Figure 1.42: *The Lancashire Independent College in College Road, Whalley Range in 1893, rear view as seen from near the junction of Clarendon Road and Churchill Avenue.*

The picture of the pastoral grounds of the college in Whalley Range, shown in Figure 1.42 is from Joseph Thompson's history of the Lancashire Independent College [2] and bears the inscription 'Meisenbach'. Georg Meisenbach invented the half tone process, enabling photographs to be printed in newspapers. He had obtained Reichspatent Nr. 22244 for it in

1882, 11 years before Thompson's book was published. The much cheaper half tone process, despite its technical inferiority decimated the use of the Woodburytype process invented by James Nasmyth's Mancunian friend Walter Bentley Woodbury in 1864.

The college had its origins in the Blackburn Independent Academy, founded in 1816 after a meeting in Manchester in the Mosely Street Congregational Chapel on the 9th of February that year. It was set up in order to educate and train young men to be priests in the Congregational Church. With few students and a realisation that Manchester, with its prodigious population growth during the 18th century had become a place where it might be viable, it moved there, having acquired a site of seven acres on which buildings costing £25,000 were erected.

The college had opened in Blackburn in 1816 with several tutors and one student. Like the Warrington Academy for Unitarians, it had to move and set up afresh in Manchester to thrive. But unlike the Manchester New College, its syllabus was much more focussed on theology and arts and the concession to science was mainly a mathematics course firmly founded on a 2,100 year old text book: Euclid's Elements.

However, when the college had been at Blackburn, John Dalton had given a course of lectures on Natural and Experimental Philosophy, which can be interpreted as most likely being on theoretical and experimental physics and chemistry. Dalton was paid £4. 18s for giving the whole course. For comparison, the records of the Manchester Court Leet for that period show that a butcher Adam Knowles was fined £5 for erecting a market stall in Shude Hill that was too wide and for leaving it for six hours whereupon 'the subjects of our Lord the King could not go return pass or ride with their horses coaches carts and carriages without great peril and danger of their lives'. Dalton's lecture course to the college was worth less than the cost of a parking fine.

The relevance of this college to physics in Manchester is therefore indirect. Eight years after the founding of the Independent College in the fields of Whalley Range village, Owens College began to teach some physics under the control of the mathematician Archibald Sandeman. It was open to both colleges to co-operate in those secular areas where there was no theological conflict. There were many links between Owens and the Independent College. Alderman Joseph Thompson, not without influence, had duties of governance at both establishments, albeit during the latter half of the 19th century. In 1844, the Independent College had noticed John Owens' large endowment for the future Owens College and remarked that it brought the day of a university nearer. By 1854, they were sending students to now established Owens to attend literary courses. There followed a series of events that Thompson relates without inhibition and with undisguised dismay, despite having been invited by the college to write the history, naming those responsible for poor governance. Thompson had been too young at the time when it had all happened to have appreciated what was going on, but was clearly shocked as he went through the papers four decades later and the pettiness, bigotry and unsuitability for office of the college governors unfolded before his eyes. As the saga unfolded, the trustees fell upon each other like starving jackals.

Thompson revealed in his history of this college [2] that the trustees of the Congregational college had clearly sought to lower academic standards by resolving to admit weaker students who were almost bound to fail the Owens courses, thus enabling the college to appoint new staff to teach 'in-house' where they had a stranglehold of control. The current students themselves, more enlightened than the governance of the college, petitioned the said governors in an articulate letter which the committee tried to suppress and have withdrawn, thus seeking to write a false account of history themselves. It is no wonder that being exposed to the perverse machinations of those who were supposed to teach them, in return for being paid, yet didn't, eventually turned students to more physical methods of protest in the next century.

The students strongly desired the link to Owens to continue and during the ensuing war of words, the true reason behind the committee's thinking emerged. The students were then essentially imprisoned within the walls of the college for five years of their deliberately extended course and were only let out to preach - under supervision. It was felt that Owens was a morally corrupting influence and the Congregational students were at risk if they entered its premises.

Out of this monstrous situation, a crack was opened in the cohort of college governors and staff, which deepened as they began to fall out among themselves following the publication of a densely printed book of over 1,000 pages written from within their midst by Samuel Davidson. It contained numerous critiques and interpretations of biblical text which were anathema to many of his colleagues. The college was rent asunder, with the participants failing to look beyond their own personal rigidity, reflecting on an individual scale, the very lack of religious cohesion which had created their dissenting church in the first place. Resignation followed resignation, including the president himself, whereafter the college's reputation was damaged for years, a fact that would not have gone unnoticed at Owens who would have quietly decided to keep the unstable college at arm's length, at least for a while. Forty years on, Thompson did not conceal his contempt for the narrow minded hardliners [2].

Although Thompson lists the subjects in the curriculum in 1825: Latin and Greek Classics, Oriental Languages, History, Geography, Mathematics, Natural Philosophy, Theory of Language, General Grammar, Mental Philosophy, Theology and Ecclesiastical History, there is no evidence whenever details are mentioned that the science teaching went beyond Euclidian geometry and algebra. At the time of their internal conflict in 1856, when the students themselves felt moved to try and influence and enlighten the arm of governance, one of the options considered by the college's committee was a recommendation to hire a professor in the literary department to teach mathematics, logic and natural philosophy. Even though Thompson's history of the college, covers the period up to 1893, the subject of 'Natural Philosophy' never appears again after 1856. To provide a full education of the sort that Manchester New College was already offering to degree level, the Independent College needed Owens. So even after the hiatus in their stability in 1854, the Independent College continued to pay Owens fees to instruct their students in subjects where they felt themselves to be deficient. Students from one college transferred to the other and *vice versa*. But closer links than this between the two colleges never developed.

Several times during the 19th century, Manchester had all the ingredients to put a strong collegiate university in place, but the Manchester New Academy, Owens and the Lancashire Independent College chose for their own individual reasons to go their separate ways. A loose federation of colleges could have been established, with each retaining its special birthright and endowments, each worshiping its own god, or not as the case might be, each having a core syllabus that satisfied its reason to exist, yet each awarding an all-embracing Manchester degree. The irony is that in order to become a university, Owens was eventually compelled against its basic wishes to form a collegiate alliance with the Leeds and Liverpool colleges, both of which did not share the Mancunian dream. The irony did not end there.

The Manchester New College eventually became a constituent of Oxford University and although the Independent College itself did not become itinerant, Thompson's history records that it sent salutations to its fellow nonconformist Mansfield College, which moved from Birmingham to Oxford in 1886. In 1889, Mansfield College finished their new buildings on a scale at least equal to the Whalley Range site in magnificence. So Oxford University, from its traditional protective Anglican theological base, was able to absorb two dissenting foundlings, albeit not fully until 1995, whereas Manchester (University) from its base of supposed freedom from a religious yoke, could not. A German word, for which there is no single English word equivalent, sums up the 19th century loss to Manchester: It was the *verschenkte* university. All of Manchester's strands of influence carry responsibility if not blame, albeit some more than others. The 'now' universities could be blamed for elitist suppression in the 17th century, but Manchester itself was to blame for the lack of academic and collegiate cohesion two centuries later.

1850: The Dream Collegiate University

Let us end this chapter with the dream sequence that never happened in reality. The three Manchester colleges (Manchester New College, The Lancashire Independent College and Owens College) and the Mechanics' Institution formed a loose alliance soon after 1851 and with the city and the North West united behind them, convinced Disraeli and Gladstone to support a new University Charter around 1868 incorporating the till then ancient and purely theological Christ's College, based in the old baronial hall buildings near the cathedral. Owens would still have been founded in 1851 in Quay Street and then built its new centre on the Oxford Road site, whilst the Manchester New College driven by growth would have felt the pressure to leave its cramped quarters in Cooper Street, moving to say, Rusholme, half way between Owens and the Independent College. The Independent College had space enough. The Mechanics' Institution might have moved further afield from Princess Street than Sackville Street and then a swathe of superb college buildings would have stretched from the Irk to the Mersey with the University able to claim its origins to be in 1422, instead of 1824 which the University now has by virtue of the beginnings of the Mechanics' Institution in that year.

Physics would almost certainly have progressed in this dream-world, much as it did in Owens, except from even more powerful foundations. Dalton would have been a professor; Joule probably not. Manchester's Christ's College, New College, Independent College, Mechanics' College and Owens College would have formed a solid and broad base on which to expand the University further in the 19th, 20th and 21st centuries. Instead, the various bodies continued as they had started, in relative isolation and some potential constituents, like the Independent College and the Congregational Church that founded it, have perished and are forgotten as if they had never been born.

Figure 1.43: *William Wylde's view of Manchester painted from Kersal Moor in 1857, six years after Owens College started.* *Attribution: Wikimedia Commons public domain.*

Of all the colleges that arose, came and went or annihilated themselves by internal bickering, the Mechanics' Institute came through its own difficult period and possibly owed its continued existence to a Swedenborgian unitarian preacher, who quite rightly took a stand against bad governance and made them change their ways. After 1851, Owens College emerged to share a common destiny with the enduring Mechanics' Institute. It became just a matter of time before the two united: 154 years no less, for the merger on equal terms. If a collegiate university had been established in the baronial village in the 17th century, could the relentless onslaught of Industry have been held at bay? Or would Manchester's collegiate Utopia have nestled in the lost pastures, which were already long buried beneath the smoky inheritance of a long lost baronial village and painted by William Wylde in 1857 (see Figure 1.43).

The reality was not the Dream World and by the end of the 19th century, only the Mechanics' Institute and Owens College, now a University, remained as a base where physics could thrive. Thus will the story now continue.

Chapter 2
1851–1870

1851: The Foundation of Owens College

In the 1850s, the *Manchester Guardian*, as it called itself until 1959, when it dropped the word *Manchester* from its title, was published only twice a week. A modest advertisement appeared in the edition of the 8th of February 1851, under the heading 'Owens College'. The advertisement announced that an institution of that name would open shortly to provide and aid the means of instructing and improving young persons of the male sex, those of an age not less than fourteen, in such branches of learning and science as are now, and may be hereafter, usually taught in the English universities. Had the copy writer been facetious, he or less likely, she, might have named them the 'now universities' except such irony might have been lost on those who had not read the Fairfax files.

On Wednesday the 12th of March, Owens College, later to become the University of Manchester, was officially launched at a public meeting in the Town Hall and an account of its opening was published in the *Guardian*. Entrance exams were held on the day after the official opening, but the college authorities thought that the heavy rain might have deterred potential students and so they put out an announcement that anyone unable to attend could sit a fresh examination the following Saturday. Thus, on Thursday the 13th, at Richard Cobden's former house in Quay Street, a 'spacious dwelling-house' (Figure 2.1 on page 115), the college opened its doors to students and, according to the newspaper reports,

> 'Shortly after 11 a.m., the staff entered, attired in collegiate gowns and carrying college caps in their hands.'

A series of inaugural lectures were then presented in the former music hall of the building, (it was such a house). Announcements in the local newspapers prior to the opening had informed the public that as well as the students who had passed the entrance examination, the first set of lectures were open to adults who presented their names at the door. The *Manchester Examiner and Times* [90], named all the distinguished adults who attended, outnumbering the students by a factor of five. The 22 youths who had registered had each paid 1 guinea admittance fee and either 3 guineas or 4 guineas course fee depending on whether they were junior or senior. The number rose to 62 by July. The college retained the whole of the admittance fee and one third of course fees, of which the remaining two thirds went to the professors, to supplement their annual salaries of £350.

All of the lectures presented on that first day were printed in full in the *Manchester Examiner and Times* article, although that by the Principal A J Scott was delayed because he had been too ill on the day to attend. The lecture by the professor of mathematics and natural philosophy, Archibald Sandeman contained virtually no physics and very little mathematics, it being a philosophical address aimed at the distinguished visitors. Nothing can be gained by reproducing the lecture here; it can be read in full in the archived copy of the above mentioned newspaper at the British Library.

Cobden's house had been acquired via advertisements placed during the last week of 1849, with professorial posts advertised in March 1850.

TO OWNERS OF BUILDINGS AND OTHERS.– The Trustees of the Fund for Educational Purposes, bequeathed by the late John Owens, Esq., are desirous of HIRING A BUILDING for the purposes of the College, to be established pursuant to the directions of his will. A spacious building, situate in an open and airy part of Manchester, containing rooms suitable for lecture and class rooms, library, chemical laboratory, housekeeper's apartments, &c., will be required. – Proposals, describing the premises and stating the rent required, and other particulars, may be forwarded to Messrs. BARLOW and ASTON, Town Hall Buildings, the solicitors of the trustees.

The trustees announced that the college would be connected to the University of London, which would allow Owens students to obtain degrees from that university, but at the same time they emphasised the importance of the establishment of a University of Manchester, able to confer its own degrees.

The first session was a short one, starting in March and ending in July thus allowing the next session to align with the norm by starting in October. By the end of that first session, there were 22 students registered for 'mathematics and physics' of which seven were juniors and 15 seniors, supplementing Sandeman's salary by £56 14s.

114

Figure 2.1:

The Owens College building pre–1851, when it was Richard Cobden's house.

The College stayed in John Cobden's former house in Quay Street for 21 years, wherafter the building re-opened as the County Court in 1878, a purpose it then served for over 100 years. The exterior double flight of steps, which can be seen in Figure 2.1, were then removed and the triangular pediment over the door replaced with a pilastered door-case. By 1900, the building appeared much as it still does today from the outside. Internally, consistent with its current (2017) use as barristers chambers, it has been completely refurbished to 21st century standards – in a style that George IV might approve of. It is a reasonable assumption therefore, that given the changes only to the door and steps in the late 19th century, that Owens college did nothing to the exterior facade when they took the building over and during the 21 years of occupancy, it looked and remained like it did when Cobden lived there.

Figure 2.2:

Piccadilly during the visit of Queen Victoria in 1851, painted by George Hayes in 1876.

Image: Courtesy and © 2017 Manchester Art Gallery.

Excitement pulsated in the city during the week that Owens opened, but not as it happens, because of the college, but rather in anticipation of the imminent visit by the Queen and her Prince Consort. The visit was in connection with the Great Exhibition of 1851. Knighthoods were bestowed and James Nasmyth was entertained in Worsley Hall, as we saw in the previous chapter. Victorian fellowships were endowed within a few years and when the college finally achieved University status, it became known as the Victoria University. Figure 2.2 on the preceding page is George Hayes' picture of the Royal visit, painted 25 years after the event.

The many driving forces behind the foundation of Owens College, most of which had been burning for centuries in Manchester, have been outlined in the first chapter. There was a strong feeling about the injustice of the obstacles placed on religious dissenters by the 'now' Universities of Oxford and Cambridge, who despite their virtual monopoly in University education in England, would not award degrees to students without a declaration of their faith to the established church. This situation did not end until the 'Universities Tests Act' was passed in 1871. There was also a tremendous need in the region, which had fuelled and driven the industrial revolution, for a thorough and relevant training in practical science.

John Owens, who bequeathed almost £100,000 for the formation of a suitable college, outlined the relevant reasons and needs with clarity:

> '... for providing and aiding the means of instructing and improving young persons of the male sex (and being of an age of not less than fourteen years) in such branches of learning and science as are usually taught in the English Universities, but subject, nevertheless, to the two following fundamental and immutable rules and conditions, that the students, professors, teachers, and others connected with the said institution, shall not be required to make any declaration as to, or to submit to any test of, their religious opinions, and that nothing shall be introduced in the matter, or mode of education or instruction in reference to any religious or theological subject which shall be reasonably offensive to the conscience of any student, or of his relations, guardians, or friends.'

This bequest, provided for the foundation of the college, translates into roughly ten million pounds by the end of the 20th century. The conditions of the bequest, which essentially declared 'No religious tests and no religious instruction' seem, at a distance of 160 years to be stated without ambiguity. This did not prevent the Manchester Church authorities and those religiously minded Owens trustees and executors from arguing

116

otherwise. The arguments delayed the opening of the college by a few years. The Church naturally wanted religious instruction of an Anglican nature in the college and went to ingenious lengths to argue their case – Since John Owens had charitably endowed the college, he must be a charitable man and hence by deduction a Christian. Logically, a Christian would want Christian instruction. Rather than indulging in a public argument, those who wanted to accomplish Owens' bequest took Mark Twain's advice to heart, never to argue in public with a fool because no-one can tell the difference. They simply published the Church's arguments in full in a local pamphlet, together with Owens' testament and let the local population draw their own conclusions.

Nevertheless, despite the clarity of Owens' bequest, it was decided to include religious instruction in the curriculum with the theologian Principal Arthur John Scott at the religious helm. Because it could not be said at the time, and then what was actually said at the time has a habit of being repeated for ever as fact, I can use the distance of over 160 years to suggest that perhaps the reason why Owens nearly failed like all Manchester colleges before it, except the Mechanics' Institute, was because Scott brought the double burden of personal theology coupled with chronic ill health to the job. This was not what Owens had in mind. Scott could not even attend the opening ceremony due to his poor health.

This initial religious decision, as discussed in Chapter 1, ensured that a collegiate expansion at university level education in Manchester in the midst of the 19th century did not happen and that the trustees of the already itinerant Manchester New College voted to continue their meandering to London instead. Yet Owens' testament declared that nothing of a religious opinion could be introduced which shall be reasonably offensive to the conscience of any student, relation, guardian or friend. In the years to follow, especially when it was moved to convert the college into a university, the government of the day repeatedly quoted Owens' testament whenever the college trustees wished to do something that the government did not like. Yet they did not follow the spirit of it themselves.

Some trustees of the dissenting Unitarian college, dissenting from the mainstream dissension, had hired the highest legal authority in the land to argue on their behalf to avoid affiliating with Owens, and yet their fears were groundless because Owens' conditions, clearly stated, would have been upheld in law. But still, the dissenting college engaged the Solicitor General to argue, successfully, that the very situation that John Owens had proscribed should not legally happen, a condition of his massive bequest, actually might. As a scientist, I am unable to reconcile this small-minded procedural obfuscation with rational thought unless there were other agendas, not stated.

I refrain from including the only extant image of Owens, a medallion, because Thompson made it clear in his history of Owens, that the medallion image did not look a bit like Owens, having been imagined from a silhouette.

Owens college eventually opened in 1851 with five professors. The college principal Scott also held the chair of logic, mental and moral philosophy and of English language and literature, a bouquet which must have cramped his visiting card. Joseph Gouge Greenwood (later to succeed Scott as principal) filled the chairs of Languages and Literatures of Greece and Rome and of Ancient and Modern History, even more of a problem for the visiting card printers. The science side had three chairs. Archibald Sandeman was the professor of mathematics. Edward Frankland was professor of chemistry and William Crawford Williamson held the chair of natural history, botany and geology. Tobias Theodores was the teacher of German, Hebrew and oriental languages. The French teacher was Monsieur August Podevin, a gentleman whose first name is never mentioned in any of the public College records. Sandeman was also required to teach any physics that might crop up, which at that time at Manchester, meant little more than mechanics. He held the entrenched view that physics was a branch of mathematics and it is stretching a point to call him a physicist. He described himself, on the title page of the books that he wrote, as 'Professor of Mathematics and Natural Philosophy.' More of Sandeman later.

Biographer Cardwell researched and read all the letters from Joule to William Thomson held by the Cambridge University Library, and in his book [15], he reported that Joule had written to his friend and scientific collaborator, William Thomson (Lord Kelvin) on the 26th of March 1850 asking if William's 28 year old elder brother James would be interested in the newly advertised post of professor of mathematics and natural philosophy at Owens. James was not interested, nor was Joule, not being a mathematician. James became professor of civil engineering at Queen's University, Belfast, the city of his birth, five years later and then succeeded the renowned William Rankine in Glasgow in 1873. Joule himself turned down a second chance in 1860.

After the college was founded in 1851, it followed the trend of most previous college enterprises in Manchester: it went rapidly downhill and almost folded. By 1858 it was in such a state of collapse, according to contemporary reports, that even the professors had noticed. About the quality of the college, the *Manchester Guardian* in an article in its edition of the 9th of July 1858 pointed out that everything had been done which it was possible for an endowed institution to do to win the young to the college:

'... but the ingenuous youth of the town resolutely shut their ears to the eloquent words of trustees and professors, and refuse to partake of the rich banquet to which they are invited. Explain it as we may, the fact is certain that this college, which eight years ago it was hoped would form the nucleus of a Manchester university, is a mortifying failure.'

The *Manchester Examiner* picked up the theme on the 20th of July:

'The worst that can be said of [the College], is that it is too good for us. It is out of place here, just as a missionary may be said to be out of his place on the coast of Africa. He offers the Gospel, and the people want Sheffield blades... The crowd rolls along Deansgate, heedless of the proximity of Plato and Aristotle... And where is poor learning all the while? Going through its diurnal martyrdom of bootless enthusiasm and empty benches.'

Even members of staff were pessimistic. Henry Enfield Roscoe, appointed to replace Frankland as Professor of Chemistry in 1857, relates [8] how he was standing one evening, preparing himself for his lecture, as he frequently did, by smoking a cigar by the back gate of the building, when a tramp accosted him and asked if the building was the Manchester Night Asylum. Roscoe replied that it was not, but if the tramp would care to call again in six months, he might find lodgings there. 62 students had enrolled in 1851 but numbers had now fallen to 33, whilst the number of professors had risen from five to six. One reason for this decline was the fact that students arriving at college from schools did not have the necessary foundation to follow the courses. To combat this local deficiency in the schools, the college started evening classes for school teachers and whether this was the sole reason or not, the corner was turned and from the nadir of 33, student numbers rose monotonically until the end of the century. Well before then, the original building in Quay Street had virtually burst at the seams: in the final year that the college occupied the building, there were almost 500 students in what was essentially nothing more than a large house, albeit one that used to have a music hall.

Following the move to new buildings of comparative spaciousness in 1872/73, student numbers rose annually to reach 1,002 by the last session of the 19th century, a remarkable progression, given the history of colleges in what was now an incorporated city. During this period, physics at the college changed from being a mathematician's hobby to being an academic powerhouse of the country. We can now consider the sequence of professors who taught physics in these early days, to see how physics in Manchester got there. It was not easy.

1851: Archibald Sandeman

Figure 2.3:

Archibald Sandeman at the time he was Professor of Mathematics at Owens with responsibility for natural philosophy.

Colour processing © 2017 Robin Marshall.

When Owens College first opened its doors in 1851, there was no department of physics – or natural philosophy – as the subject was called in those days. Any physics teaching was in the hands of the professor of mathematics, Archibald Sandeman. Despite his strongly held views on the teaching of physics, there is not much evidence that he actually taught any. However, he did pioneer the teaching of the logic of mathematics and wrote a book about it called *Pelicotetics* [91]. Thomas Barker succeeded Sandeman as professor of mathematics in 1865, and J J Thomson said that he never knew a better teacher of mathematics than Barker. Along with all other students, Thomson did find Sandeman's book difficult, 'page after page without a full stop and none who started it ever got far.' This book by Sandeman was published in 1868, by which time Sandeman gave his location as Queens' College Cambridge. Thomson did not say whether the book was recommended reading or that he sought it out himself.

Sandeman had already been appointed to his chair well before the college opened in 1851. On the other hand, the competition for the chair of Chemistry had been fierce and was eventually secured, just in time, by Edward Frankland (see Figure 2.4) who received eleven glowing letters of reference, including two from Wilhelm Bunsen and Justus von Liebig.

Figure 2.4:

Edward Frankland. The first Professor of Chemistry at Owens College and the discoverer of valency.

Colour processing © 2017 Robin Marshall.

Charles Dickens noted Frankland's appointment in his regular journal *Household Narrative of Current Events*, published between 1850 and 1859 [92]:

In this journal, Dickens kept his finger on matters concerning Britain, the Empire and when it mattered, the rest of the World. This was the only occasion in the nine years of the journal's existence when he mentioned Owens College.

Writing on *Science in Victorian Manchester* [96], Robert Kargon in 1977 could find no obvious reasons why the chair of mathematics went to the relatively obscure Sandeman.

In comparison, Frankland shines as a beacon of discovery. His deserved reputation as one of the most foremost 19th century chemists was suppressed by his peers, mainly because he was born illegitimate in a Victorian age and this was enough to obscure his achievements and make him an easy target.

Even as late as 1906, the lack of peer recognition was still causing bewilderment in the relatively new State of Germany. In his book *A History of Chemistry* in 1906, Ernst von Meyer [99] attacked the bias of Wurtz's *Histoire de Doctrines Chimiques* as well as Kekule's textbook on Organic Chemistry:

'It seems hardly credible that Frankland, the real originator of the doctrine of valency, should never be mentioned in this publication. The same applies to the general section of Kekule's *Lehrbuch der organischen Chemie*; there the debt due to Frankland is absolutely ignored, while the share in the development of organic chemistry taken by Dumas, Gerhardt, Laurent and Kekule himself is minutely detailed.'

Wurtz later acknowledged Frankland's singular role in the concept of valency. Many apologised for Wurtz, saying that he had been consumed by a nationalistic and Prussian agenda rather than moral one. 'La chimie est une science Français; elle fut constitute par Lavoisier.'

Many could argue that this is no excuse. Indeed, the subsequent Franco–Prussian war of 1870, the Schleswig-Holstein problem, WW1 and WW2 were all consequences of a nationalistic and Prussian agenda with which valency ought not to be tarnished. Yet although Wurtz's sentiments have been apologised for by many, claiming that Wurtz had other things in mind, it remains an irrefutable fact, that as history unfolds, written words rather than the unspoken qualifying thoughts behind them are all that remain centuries on.

122

Joseph Thompson, the chronicler of Owens until it became a university, described how Sandeman came to be ousted [1]. During the early years of Owens College, when it floundered, but still managed to escape the demise that engulfed all other ambitious colleges in Manchester except for the Mechanics' Institute, Roscoe found himself increasingly teaching physics as well as chemistry. For Sandeman, the final blow came in 1858 when the University of London, whose degrees were awarded to Owens students, introduced the degrees of BSc and DSc, requiring a deeper knowledge of physics than was being taught at Owens. Student numbers at Owens had just started to rise and it was against this local trend, that the trustees felt that Owens College was deficient in this area. Therefore, the trustees felt the great importance of being fully equipped if they were to make the college successful in its scientific teaching. Equipped when used in this sense, as it was used by Joseph Thompson, who wrote the history of the college, meant having the right team of scientific teachers and researchers. Roscoe saw and seized an opportunity to orchestrate Sandeman's removal, cleverly arranging for Augustus de Morgan (a most prominent mathematician) and Sir George Stokes (Lucasian Professor of Mathematics at Cambridge) to submit written arguments for or against appointing a dedicated physics professor.

At a college meeting held on the 23rd of January 1860, the Principal of the College stated that the trustees wanted to have the collective opinion of the professors as to the desirability or otherwise of creating a professorship of natural philosophy. Up till then, he said, the subject had been included within mathematics. Natural philosophy was now commanding greater attention than ever before, and London University had made it one of the special subjects for its new degrees.

Roscoe described [8] how it was he who brought this proposal before his colleagues through the new Principal Greenwood. From his German experience, Roscoe knew what a course of experimental physics should be and asserted that there was no such course in the college apart from the elementary bits of physics he covered in his own chemistry course. Sandeman, he politely remarked, was the purist of pure mathematicians, and had not the slightest idea of experimentation. Roscoe believed that Sandeman's aforementioned book *Peliotetics* was an attempt to prove why two plus two equals four and not five.

The professors promptly passed a resolution that great inconvenience had long been felt from the absence of a course of lectures in the college on experimental physics. This was a splendid resolution although what little evidence exists suggests that physics was not high on most peoples' agenda in the mid 19th century. The course referred to was expressly required in preparation for the new matriculation examination of the University of London to which Owens subscribed. It was also very desirable as collateral to the chemistry courses. Bearing in mind the greater prominence given to natural philosophy in the new regulations for the degree of BA and in the institution of degrees in science, the college now felt unable to offer to its students, complete courses of study conducting to those degrees in arts and science, preparation for which might reasonably be sought in a college offering general education. The meeting of College professors therefore thought that a professorship of natural philosophy ought to be created and that the professor should not be a mathematician. Archibald Sandeman, professor of mathematics and responsible for teaching what little physics there was on the curriculum, dissented from this view. His reasons were unfolded at a later meeting when the views of two extremely eminent external mathematicians were also considered. The college principal was asked by the meeting of professors to submit some letters which the College trustees had received from the two distinguished professors of mathematics, de Morgan and Stokes, in reply to letters that Roscoe had written to them. Their views had been sought, according to Joseph Thompson:

> '... both as the convictions of eminent men, and as the pleadings for a separate course of study which was strongly opposed by some in those days, but which had now become generally accepted.'

Roscoe's enquiring letter to Stokes (quoted in Cardwell [15]) contained the following:

> 'Is it advisable to have two courses of lectures, one an experimental course & one a course in which the subject is treated more mathematically & for which the student should be required to show a certain amount of previous mathematical training? What degree of mathematical training should be considered necessary?
>
> Then I should be glad to hear what sum of money you think would be required to purchase a tolerably complete physical cabinet containing such apparatus as is required for illustrating a course of lectures. Do you think that with an endowment of £200 per annum with the addition of fees to the amount of £60 to £100 we could secure the services of a good man! Is there any way by which a man holding such an office might supplement his income in a place like Manchester?
>
> I understand fully what you said about the difficulty of obtaining a man great in both experiment & in mathematics. We, in England,

have, as you say, no school for physicists & I fear that a mere mathematician who has only book knowledge of the subject would answer but badly for us. Do you know of any young Cambridge men who might possibly be suitable for the post? I have accidentally heard that a fellow of St John's, Clifton by name, takes a great interest in such pursuits – are you acquainted with him?

Of course one's first idea in Manchester was respecting Joule – but I doubt if he would accept the office. At any rate the selection will be an open one.'

There was hardly a more eminent mathematical physicist in the land than George Gabriel Stokes, F.R.S., Lucasian Professor of Mathematics at the University of Cambridge, a post once held by Newton and subsequently by Hawking. He is pictured in Figure 2.5.

Figure 2.5: *Left: George Gabriel Stokes and Right: Augustus de Morgan. The image of Stokes is derived from the frontispiece of his autobiography and that of de Morgan from his visiting card.*
Colour processing © 2017 Robin Marshall.

Roscoe probably also wrote a similar letter to the mathematician de Morgan (also shown in Figure 2.5). Cardwell points out that the term 'mere mathematician' was a somewhat tactless phrase to say to a Lucasian Professor of Mathematics, to which I would counter that

Lucasian professors such as Newton have always been far more than mere mathematicians and probably both Roscoe and Stokes knew that.

Stokes wrote back [1] – in the canonical single paragraph of the age:

'I entertain a very strong opinion as to the great value of a course of lectures, mainly experimental, on natural philosophy. I think it a great defect in our system here, that our students have so little opportunity for attending or encouragement to attend lectures of this kind. The study of natural philosophy, for its own sake and not merely as a field for the exercise of mathematics, is, I think, too much neglected among us. The consequence I believe to be that, besides the loss of a particular and valuable kind of mental training which the study of physics affords, where a class of faculties is called into exercise allied to but distinct from the faculties exercised in the study of mathematics, some even of our good mathematicians are very ill acquainted with practical physics. I think that in any establishment for higher teaching there is ample room for a chair of natural philosophy, distinct from one of mathematics, at least if either the number of students be large or mathematics be taught to a great extent. If the two offices be combined in one man, there is great danger either that his taste for physics and the time he must devote to the preparation of a good course of experimental lectures will lead him off from mathematics, or else his taste for mathematics will lead him off from experimental physics. I think a lecturer on physics ought certainly to know mathematics and ought freely to use them when occasion arises, and his class ought readily to understand simple geometry and algebra. Without that there is danger that his lectures may merge in mere vague expressions of popular notions altogether wanting in precision. But, on the other hand, a course of lectures on physics would be utterly spoiled if the lecturer merely used physical laws as pegs whereon to hang mathematical problems instead of seeking to investigate and establish them for their own sake, merely using now and then a little mathematics as a tool. I think it is not very common to meet with men who have a strong taste for both mathematics and physics, and even if you were to meet with such a man, there would be work enough in a large establishment for a mathematician and a physicist.'

Augustus de Morgan was professor of mathematics at University College, London and a complete master of logic, also more than a 'mere mathematician.' de Morgan wrote a substantial and well reasoned letter of support for physics as an independent subject, albeit in the prevailing Victorian style of a single long paragraph. In his letter he said:

'A course of experimental physics *may* be as rigorous in its way as a course of mathematics. It may be *ocular demonstration* of the consequences of admitted principles, or *the same* of the principles themselves, or both. This depends on the teacher. An illogical man may make mathematics as unsound as what is popularly called physics in lectures: a logical man may make experiment as soundly demonstrative of its conclusion as mathematics itself. But he must know what experiment can end in, and what must be prepared to prevent his pupils from imagining that they do demonstrate what they really do not. If the college can find out the man, the man will find out the way. But he must be a sound mathematician. I never knew of a physical lecturer who was no mathematician who understood what it is that experiment does teach. Take a mathematician who has to learn his work rather than a ready made experimentalist without mathematics. I hold that a demonstrative course of experimental physics is in itself desirable. It shows what *demonstration* is out of mathematics. And, more important still, in proper hands it shows what demonstration is not... The fault of lectures in experiment is that they aim at too much detail. Method is to be taught and quantity. When the student has been on fundamental points under a sound instructor, he *can read experiment* with profit; without such previous teaching, he can make nothing of his reading. It is anatomy without dissection and conchology without any shells. This is always forgotten almost. The lecturers think that they must show everything. They ought to show as much as will make a good book show the rest... I am sure that fifty lectures would be ample for the fundamental teaching. The object is not to show the practical application of science so much as to put the student in a condition to take the showing in the places where it is shown. I will defy any one to show the practical application of science in college lectures.'

It was thus a matter of pragmatic deliberation by Roscoe, rather than chance, that the College authorities sought guidance on the relationship between the teaching of mathematics and physics from the two most distinguished mathematicians in the land and did not ask any physicists to comment, since such a set of physicists would have most likely included the very person they were likely to hire. The nature of their comments ensured that the decision was cast in iron and nothing that Sandeman, a far lesser mathematician, could say, would alter the fact. This did not prevent Sandeman from trying to cling on to physics teaching, which he wasn't doing anyway, as Roscoe claims. At the subsequent meeting of the trustees where they considered what to do, Sandeman read a paper outlining his dissension to the proposed course of action.

Alderman Joseph Thompson, who was present at this (and almost all other meetings involving the college) described Sandeman's contribution, quoted below, as a 'rhapsody on mathematics':

'The mind found in notions of time, number, space, quantity, and motion, so much its own, that it set to work upon them to weave out their general relations all by itself, without needing to return to the objects which called them forth. The mind, indeed, not only thinks these relations but over and above finds them to be so altogether part and parcel of its very self that it can think no otherwise. In natural philosophy the mind has to play no less important, no less needful, and not so exclusive a part as in mathematics; and

(1) asking questions
(2) understanding the answers,
and
(3) following out the consequences,

geometry and mathematics generally became a necessity to this end. Of the many ways in which this necessity showed itself the three following might be noticed:

1st. Each kind of magnitude handled in natural philosophy had something which belonged to itself alone, and something which it had in common with other and even with all kinds of magnitudes...

2nd. The laws of natural philosophy being laws of magnitude could only be stated in mathematical form...

3rd. Laws of any generality could but be suggested by experiments, however many, varied, or exact.

A general law must after all be in the first instance by some happy thought laid hold of, and can only be established by its consequences being found true. These consequences can be traced by no hand but by the mathematicians. It was precisely in the darkest and most outlying regions of natural philosophy that mathematics was most a stranger, just where indeed anything like philosophy was as unknown as mathematics. Mathematics gets some of the most striking and beautiful illustrations of her truths at the hands of natural philosophy. Indeed, the two were entwined together in most loving, helpful, sisterly bonds, and some of either's fairest and richest provinces were the other's gift.'

These words, glistening with 'midnight oil' can only be admired, but from a distance of one and a half centuries, they can be read equally as an argument why mathematics, being an essential tool of physics, ought to be under the wing of a physicist. They are the floundering words of an amiable mathematician as he saw physics drift inexorably from his grasp.

Similar arguments could also be made about the relationship between physics and engineering, but never are. Both arguments are equally fallacious. A mirror of these circumstances occurred in Edinburgh many years later when the emigrant theoretical physicist Max Born was appointed to the chair of applied mathematics. He objected strongly to the name 'applied mathematics', preferring the more relevant 'theoretical physics' and once he was installed in Edinburgh, he set about changing the name. He used the argument that experimental physics at the time relied on a substantial amount of glass blowing, perhaps as much as theoretical physicists relied on mathematics. No one at Cambridge or Manchester, Born argued, would have contemplated calling their physical laboratory by the name 'Department of Applied Glass Blowing'.

Stokes and de Morgan had written their letters of strong support for physics to Manchester in 1860 and their opinions had an immediate effect in Manchester. Yet strikingly, it was not until 1869 that a specially appointed committee of Senate at Stokes' own University of Cambridge, recommended the 'founding of a special professorship (of experimental physics) and of supplying the professor with the means of making his teaching practical – in other words, of giving him a demonstrator, a lecture room, a laboratory and several classrooms, with sufficient stock of apparatus.' Even then, neither the University nor the separate colleges at Cambridge financed the proposed scheme. The University could not afford it and the colleges which might have been able to afford it, would not. It was left to the 7th Duke of Devonshire in 1870 to endow a new scientific laboratory at Cambridge, costing £8,450 (equivalent to about half a million pounds by the end of the 20th century). The Duke, whose maternal great-grandmother, Elizabeth Cavendish and paternal great-grandfather, the 4th Duke William Cavendish, were siblings, was both a senior wrangler and a winner of the Smith's prize for examination performance in 1829, an honour he shared with subsequent prizewinners, John Herschel, Stokes himself, Arthur Cayley, William Thomson (Lord Kelvin), James Clerk Maxwell, Horace Lamb (Professor of Mathematics at Manchester 1885–1920), Joseph Larmor and J J Thomson. In time, the prize became awarded for an essay, and subsequent winners include Arthur Stanley Eddington (Chapter 4), Alan Turing (Chapter 7) and Fred Hoyle. The scientist Henry Cavendish, of whom we shall hear more in Chapter 3, was the grandson of William Cavendish, the 2nd Duke.

Thus, in 1860, however, we can recognise a turning point in science where, provoked by a progressive decision at University College London, physics in Manchester emerged as a distinct subject. It meant that any scientist proficient in experimental techniques and with a deep insight, could be a leader and professor of physics in his own right. He did not have to be trained initially as a mathematician. At their next meeting on the 29th of March 1860, the Owens College trustees lost no time in passing a resolution establishing a professorship of natural philosophy, and they referred it to a subcommittee to consider the general nature and limits of the professorship, and the steps to be taken towards the appointment of a professor. The new professor, Robert Bellamy Clifton was found, hired and was on his chair within four months.

Sandeman might have thrived in a modern larger department where his eccentricity would have interacted provocatively and positively with conservative colleagues. He continued as the professor of mathematics without a physics portfolio until 1865 when he was quietly replaced by the more eminent mathematician Barker. Kargon [96] remarks that there were too many negative stories about Sandeman's teaching for there to be no foundation. There are many 19th century textbooks that one can pick up now and admire the depth of their content. Preston's textbook on Heat, one of Henry Moseley's inscribed school prizes, now in the University archives is one such book. Therefore it would not come amiss to review Sandeman's book today. If Euclid's books stand up to scrutiny today more than 2000 years after being written, then Sandeman's ought to do so as well. The full title of the book that J J Thomson was forced to read is *Pelicotetics or the Science of Quantity An Elementary Treatise on Algebra and Its Groundwork Arithmetic.* Sandeman asserted the purpose of his book in the opening paragraph of his Preface:

'This book seeks to make Arithmetic and Algebra a science, – a piece of knowledge to wit everywhere reasoned out in an orderly way from principles expressly laid down –, and toward that end has to run wide of the track of the common books.'

The final paragraph of his Preface, however, shows more of the man than the mathematics:

'Small need then to say as a wind up that arithmetic and algebra in their wonted setting forth cannot but be educationally bad and mischievous scientifically misleading bewildering unhelping balking stunning deadening and killing and philosophically worthless.'

Here, the term 'wind up' meant that he was closing or winding up the preface. Any student, whether one as sharp as J J Thomson or one of his contemporaries more prone to breaking thermometers (see Chapter 3), could not be otherwise but provoked to read on. There is a section towards the end of the book: 'Numerical expression of the circumference of a circle in reference to the diameter as unit.' which gives a glimpse of why Thomson said what he did about this book. With reference to Figure 2.6, here is the first prodigious *sentence* from that section, Sandeman's method for calculating π. The punctuation in this *paragraph* is representative of the whole book, where one may search in vain for a comma or full stop.

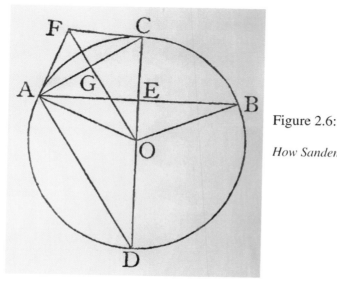

Figure 2.6:

How Sandeman calculated π.

'Let *AB* be a side of a regular polygon inscribed in a circle *ABC* by drawing the chords of more than two equal arcs into which the circumference is wholly cut through the circle's center *O* which is not in *AB* because *AB* is not a diameter and *AB*'s middle point *E* which is within the circle because in between the ends of *AB* draw a straight line and let *CD* be the several points where this straight line cuts the circumference on the opposite side of *E* to *O* and on the same side of *E* as *O* this straight line *CEOD* too because it meets and does not lie wholly together with *AB* meets at *E* only and there cuts *AB* so that *D* is on the same side of *AB* as *O* and *C* on the opposite side it is therefore the arc *ACB* which is one of the equal arcs making up the whole circumference and the arc *ABB* which is made up of all the rest join *OA OB* since *O* is not in *AB OEA OEB* are triangles and *OEA OEB* have the two sides *EO OA* equal severally to the two

131

EO OB and the bases *EA EB* equal therefore the angles *EOA FOB* are equal and therefore the arcs *CA CB* are equal whereon these equal angles at the center stand if then the chord *AC* be drawn and *n* be the number of equal arcs that make up the whole circumference or the number of sides of the inscribed regular polygon by bisecting each of the *n* − 1 other arcs as well as *ACB* the halves of the *n* equal arcs are all equal and the circumference is thus wholly cut into *n* × 2 equal arcs hence *AC* is one side of a regular *n* × 2 sided polygon that may be inscribed in *ABC* by drawing the chords of the *n* × 2 arcs and two straight lines *AF CF* may be drawn touching *ABC* at *A C* shutting in with the chord *AC* an isosceles triangle *FAC* upon the opposite side of the chord *AC* to *O* and each a half side of a regular *n* × 2 sided polygon that may be circumscribed about *ABC* with the sides touching *ABC* at the *n* × 2 arc ends.'

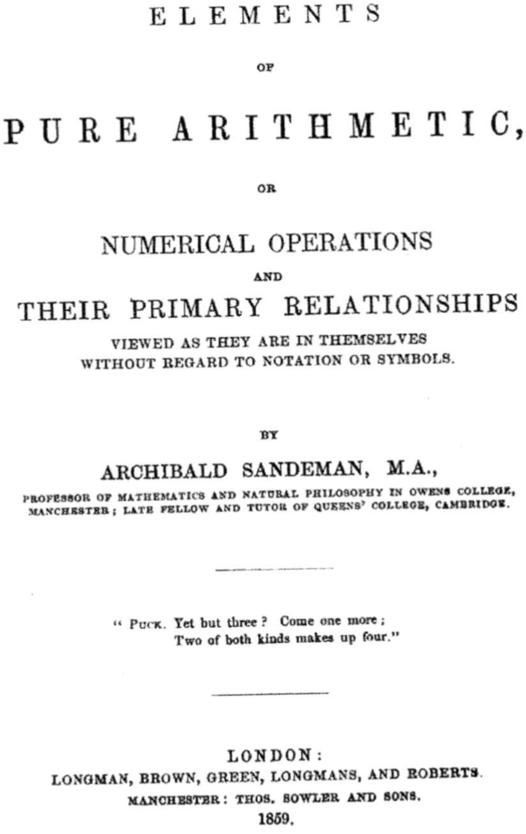

ELEMENTS

OF

PURE ARITHMETIC,

OR

NUMERICAL OPERATIONS

AND

THEIR PRIMARY RELATIONSHIPS

VIEWED AS THEY ARE IN THEMSELVES
WITHOUT REGARD TO NOTATION OR SYMBOLS.

BY

ARCHIBALD SANDEMAN, M.A.,

PROFESSOR OF MATHEMATICS AND NATURAL PHILOSOPHY IN OWENS COLLEGE,
MANCHESTER; LATE FELLOW AND TUTOR OF QUEENS' COLLEGE, CAMBRIDGE.

———————

" PUCK. Yet but three ? Come one more ;
Two of both kinds makes up four."

———————

LONDON:
LONGMAN, BROWN, GREEN, LONGMANS, AND ROBERTS.
MANCHESTER: THOS. SOWLER AND SONS.
1859.

Figure 2.7:

Sandeman's book on Pure Arithmetic, published whilst Professor of Mathematics and Natural Philosophy at Manchester.

This was not Sandeman's only book. In 1859, a year before he was deposed as the professor responsible for physics at Owens College, he published a book entitled *Elements of Pure Arithmetic* (see Figure 2.7 on the preceding page). The title page states that he was 'Professor of Mathematics and Natural Philosophy in Owens College, Manchester; Late Fellow and Tutor of Queens' College, Cambridge.'

The title page contains a quotation from Midsummer Night's Dream: 'PUCK: Yet but three? Come on more; Two of both kinds makes up four.' This is more likely to be the source of Roscoe's above quoted comment on the *Pelicotetics* book. Sandeman is shown in the physics department records as having resigned in 1860 although he continued as Professor of Mathematics until 1865. His resignation as Professor of Natural Philosophy in 1860 is therefore a mere technicality in his career. Kargon [96], writing in modern times, claimed that in general, Sandeman continued to be regarded as a detriment to Manchester's reputation with too many Sandeman stories to ignore, which undermined his effectiveness as a teacher. The sources of these stories are not cited. Moreover, buried in the *Annual Report of the Council to the Court of Governors* of the new university, submitted in October 1881, is a discussion on Sandeman's mathematics teaching skills 30 years previously, cleverly avoiding his physics teaching skills:

> 'But a still more striking illustration of this process will be found in the history of the chair then (1851) filled by Mr. Arch. Sandeman. He was styled Professor of Mathematics and Physics, and all who remember him will know how ably and with what originality he taught the several branches of Pure and Applied Mathematics.'

The discrepancy between this generous view and Roscoe's view and actions can partly be explained by the passage of time. But another likely element is that Roscoe, who by now had become a Member of Parliament, was inevitably a politician and his statements on other scientists were motivated by his own agenda, even friendship, but in this case having to teach physics when it was someone else's job. He eulogised on Sandeman's successors as professors of natural philosophy – Robert Bellamy Clifton and William Jack. Clifton never did a scrap of research and mercifully left after six years to become Head of Department at Oxford for over 40 years during which time he thwarted any attempts by others to do original research. William Jack was even less of a physicist than Sandeman and fled the physics chair in Manchester to become the editor of the Glasgow Herald, which according to Jack's sympathetic biographer, Adolphus Ward [104], 'had soon fallen on troublous times'.

Yet Archibald Sandeman cannot be faulted for his public spiritedness. Upon his death (26th of June 1893) a sum of £32,000, massively generous, was bequeathed to establish a public library in Perth and Kinross, which subsequently bore his name. A photographic portrait of Sandeman is listed by Hartog [4] among the art and photographs owned by The University of Manchester in 1901, but it was marked as 'missing' during a 1975 stock taking. The portrait reproduced above (Figure 2.3 on page 120), is based on a digital copy of a black and white photograph kindly provided by the (now renamed) library that Sandeman endowed. I present here nothing more than theoretical speculation, but the Perth and Kinross library and gallery owns an oil painting of Sandeman which was painted by Charles Andrew Sellar in 1922 using a photograph as the basis. The artist died not long after in 1926. It is not beyond the realms of possibility that Sellar acquired the photograph from the University of Manchester on loan and died before the photograph was returned. This is pure hypothesis.

Archibald Sandeman takes his place, whatever the rumours, as the first in a long line of mathematical physicists in Manchester, perhaps more mathematician than physicist.

1854: A New Building for the Mechanics' Institution

The Manchester New College, fatally floundering in 1851 and choosing not to merge with Owens, relocated to London. The Mechanics' Institution, a short walk from Owens College in Quay Street, was thriving and in 1854, the original building in Cooper Street was vacated, the Institution moving into splendid custom built premises in David Street. Traveling out of town, Princess Street became Bond Street which then became David Street. Now they are all called Princess Street.

The 1854 Institute building, designed by local architect J E Gregan, stands today as a Listed Building and was already used in 1868 as the venue for the first meeting of the Trades Union Congress and is still nurtured by the TUC. Figure 2.8 on the next page shows a contemporary drawing from 1854 and a photograph I took 156 years later.

1858: Owens and the Mechanics' Institution joint project

In January 1858, a working man's college was inaugurated at the Mechanics' Institution with the support of Owens College, a joint enterprise that presaged the eventual union of the two bodies as the largest

Figure 2.8: *Left: The Mechanics' Institution Building in David Street (now 103 Princess Street), Manchester in 1854. Right: The Former Mechanics' Institution Building, 103 Princess Street, Manchester in 2010. The building is now nurtured by the Trades Union Congress.*
Photo: Robin Marshall.

campus university in the land. The management committee contained notables from both bodies as well as the London born, Salford Alderman Edward Ryley Langworthy. A circular was issued stressing the collegiate nature of the enterprise, bringing the definition of a college into more general use:

'A *college* implies that its members are associated by an interest in some common aim; that not the advantage of the individual alone is dear to each, but that of the whole society; and that this common good is furthered by a hearty fellow-feeling among the members.'

Several subjects were offered, and a course of ten one hour lectures after working hours, between 7:30 and 9:30 pm cost two shillings. For this, the working students, who had to be sixteen years or older received instruction from the best of both Owens and the Mechanics' Institution. Sandeman taught arithmetic and algebra, whereas physics was offered in the form of the principles of mechanics by the Rev Alfred Newth. Newth was the professor of Hebrew and philosophy at the Lancashire Independent College in Whalley Range. It is hard to imagine a Manchester physics professor failing to die of shame if lectures on physics were presented in the city by a professor of Hebrew and philosophy and not the physics professor himself. But Sandeman was no physics professor. Owens' Principal Scott taught political philosophy and the future Principal Greenwood taught Latin and gave one free course – a bible class. Any notion that Owens was a completely secular execution of Owens' legacy does not stand up to scrutiny in any age. Bible classes were inserted whenever possible. The secular origins of Owens College never stand up to close scrutiny. Many wanted it to have a religious core and the history of the previous century shows that if they had had their way, the college would have failed miserably. It would have sunk, inevitably, into an abyss of internal bickering.

A People's College had been operated successfully in Sheffield by the Rev R S Bayley since 1842 and despite doubts about propriety of such heresy, women were also admitted, it being found by experience that the admission of women increased decorum and self-respect among both sexes. The scheme was copied in London and then by Manchester. The Sheffield college held classes between 6:30 and 7:30 in the morning and 7:30 and 9:30 in the evening.

But after three years, problems arose within this joint enterprise over room space and funding and the scheme was absorbed into Owens' regular evening classes, which taught more students than those who enrolled for daytime classes. Indeed, the evening classes, attracting several hundred students, were making Owens College financially viable. Before the demise of Sandeman and the introduction of a formal physics course at Owens, elementary physics was also being taught as part of a science syllabus to the working men of the city by the professors at Owens and the Independent College.

1860: Robert Bellamy Clifton

After the sequence of events leading up to the absolvment of Archibald Sandeman from teaching any physics, Robert Bellamy Clifton was appointed to the new chair of Natural Philosophy on the 31st of July 1860, thus increasing the size of Senate from seven to eight. Roscoe [8], described Clifton as a 'distinguished young Cambridge man' at the time of his appointment, and within a short time at Owens as: '... most popular, his lectures were most admirable, and enabled me to dispense with teaching any portion of the subject'. Thus we hear from Roscoe that Sandeman was happy to leave the teaching of physics to chemists but could not come to terms with the appointment of a professor in the subject.

Not much about Clifton remains on record within the University apart from a portrait from his Oxford days (Figure 2.9 on the facing page) and the 1863 Senate montage (see Figure 2.10 on the next page). Sandeman, second left in the Senate montage, was still in post as professor of mathematics. Roscoe, Sandeman and Scott are clearly recognisable from other contemporary photographs so we can take it that the picture of the

26 year old Clifton, of whom no other image exists from his youth, is also a good likeness.

Figure 2.9:

Robert Bellamy Clifton, Professor of Natural Philosophy at Owens College, 1860–1866.

Colour processing © 2017 Robin Marshall.

Figure 2.10: *The Owens College Senate 1862–1863. Left to right: H E Roscoe, A Sandeman, R C Christie, R B Clifton, J G Greenwood, A J Scott, T Theodores, W C Williamson.*

Teaching was always Clifton's forte. It remained so for the whole of his long career, which fortunately for Manchester was carried out elsewhere. He regularly attended the meetings of the 'Lit. and Phil.' and from their published journals, his rather limited research activities can be established.

Soon after Clifton's arrival in Manchester, the city was overwhelmed by events happening thousands of miles away. The developments in the American civil war meant that the regular cotton supplies from the 'Deep South' effectively dried up. Factories in Manchester, on which much of the local economy was based, came to a standstill and there was massive unemployment in the area. According to Roscoe:

> 'There was some danger of a depression of spirits occurring, which might lead to serious results if the attention of the unemployed was not turned in some new direction.'

In 1862, a suitable 'new direction' was chemistry and physics lectures in the evenings. Disused mills and other large rooms were hired and a series of entertainment was started including lectures on various topics including chemistry and physics. Roscoe claims an average weekly audience totalling 4,000. Initially, Roscoe would lecture on the chemistry of a candle and Clifton lectured on the physics of heat.

Roscoe then set up a series of 'penny lectures', claiming that they were partly as a consequence of these 'recreation evenings', and partly because of the appreciation of science which was shown when the subject was treated popularly. The lectures were printed up and published by John Heywood of Deansgate and sold for a penny each.

About 50 men attended during the first winter. Soon, public lectures on science were being held as a course of 13 lectures for 2s 6d. The lectures were extended to include eminent guest speakers and there were occasions when the Free Trade Hall and Hulme Town Hall were filled to capacity. Roscoe asserts that 3,700 people were present at the lecture by John Tyndall on *Crystalline and Molecular Forces.* in the Free Trade Hall and subsequent lectures drew an average of 675. The series of lectures by eminent visitors were published as pamphlets and eventually in the form of a bound volume entitled *Science Lectures for the People.* The book, which I proudly own, has a superb cover, as shown in Figure 2.11 on the facing page. Despite the local poverty, a few merchants in Manchester became cotton millionaires overnight because they held stocks of the now rare commodity. Sympathy for those who mercilessly exploited their temporary advantage was hard to find when the Civil war ended and cotton prices collapsed.

Figure 2.11:

The cover of the 1873–74 edition of The Manchester Science Lectures.

Clifton occasionally made a presentation during his regular attendance at the Lit. and Phil. meetings, although he usually did not have much to say. Sometimes he did not write up the presentation as a paper, leaving a sub-editor to produce a brief summary. Thus at the meeting of the Mathematical and Physical Section, on the 5th of March 1863, having just been appointed secretary and possibly writing the summary himself:

'Professor Clifton exhibited an instrument by him for observing the phenomena of Conical Refraction, both internal and external.'

On the 10th of November 1864, he exhibited an acoustical electric telegraph, by which a note, sounded at one end of the line, was reproduced at the other. He drew attention to the fact that MM Bourget and Bernard had shown, in agreement with the mathematical investigations of Poisson and M Lamé, that a given square membrane will not vibrate in accord with *any note* and so probably, his circular membrane would not either. He did not investigate whether it would or would not. The development of the telephone can be traced from the first tentative steps in 1833 to a successful patent in 1876. Already in 1860, the German Johann Philipp Reis had transmitted musical notes and speech along a wire to a receiver. Instead of saying 'Komm mal hier Herr Watson, ich will dich sehen', Reis uttered words he thought might be more distinctly heard: 'Das Pferd frisst keinen

Gurkensalat' (Horses don't eat cucumber salad.) The German post office issued a commemorative 80 pf stamp to Reis on his 150th birthday in 1984 on which an idealised image shows Reis holding a 'practical' telephone. Of course Alexander Bell did not invent the telephone any more than Thomas Edison invented the light bulb. Both got patents for their specific devices despite decades and even centuries of prior work by others.

The relevance here is that Clifton gave a primitive demonstration, not much more than at schoolboy level and there is no evidence that he did any serious research into the subject except how to give an entertaining talk. The contributions from Clifton to the Lit. and Phil. are unimpressive when viewed alongside those of his contemporary Joule. Given that Joule's friend William Thomson/Lord Kelvin sometimes attended the meetings, there may have been embarrassment. However, Thomson did give a generous reference when Clifton successfully applied for the Oxford chair in 1866, as did all the other referees who knew him well.

Clifton addressed a matter of some importance on the 14th of November 1865 when he presented a paper which dealt in more detail with a subject he had raised on the 5th of March 1863, on the same day that he showed his conical refraction device. He attempted to understand the production and emission of light from bodies either by incandescence or fluorescence. The spectral lines of luminous elements were well established and no one of course knew how to interpret the various line frequencies in terms of a model of the atom. Frequencies meant vibrations. But exactly what was vibrating was not known by anyone at the time. Clifton sought the answer. Unfortunately, like so many of his contemporaries, he got bogged down in the aether. It seems strange with hindsight, to see a physicist assign untested properties to a medium whose existence had never been proved, in order to reach new conclusions which therefore could never be proved. This is the sort of stuff to incite philosopher Thomas Reid. Possibly Clifton, like so many others, sought to establish the explanation of a phenomenon, like spectral lines, that was not understood and at the same time to absolutely prove the existence of the aether. Clifton cannot be criticised for being a man of the aether, everyone else was at the time. There is an almost direct parallel today. Strings, or superstrings have become very fashionable theoretically as the latest 'fundamental' layer of material structure in the universe. These tiny, but exceedingly stiff loops would have a size of about $10^{-33} - 10^{-35}$ of a metre. The smallest dimension considered worthy enough to be given a name is the yoctometre, a prodigiously tiny 10^{-24} m, eleven orders of magnitude (one hundred billion times) bigger than a string. I have reserved the name planckometre for 10^{-36} m, although this probably violates a strict naming rule that I am not aware of. Even a yoctometre itself is six orders of magnitude, a million times smaller than the tiniest resolved size, established in particle physics experiments. In deference to string advocates, it should be recognised that they aim to show that strings can reach out and have effects on the scale of millimetres. This is a reasonable assertion because the effects of particles on the scale of 10^{-18} m are able to reach out and affect the behaviour of the Universe on a scale of 10^{26} m. But time alone will tell whether or not strings are the phlogiston or aether of the 21st century.

Until technology advanced, and the speed of light could be measured with sufficient accuracy, the presence or otherwise of the aether could not be proved and until it was dismissed, its hypothesis spawned a host of unlikely consequences. Clifton was not alone in being swept along by the aether and should not be blamed. Others wallowed in deeper misconceptions. Without knowledge that the electron existed, a vortex model of the atoms had arisen and well into the 20th century, even after the discovery of the electron in 1897, school and primary undergraduate textbooks were still carrying pictures of proposed models of the atom based on vortex rings. Figure 2.12 is copied from the 1902 edition of Preston's textbook of heat, one of Henry Moseley's book prizes from his school (Eton), now in the John Rylands Library archive.

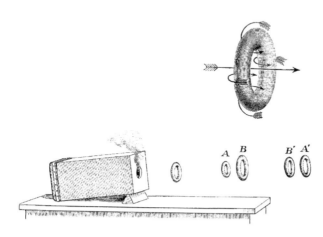

Figure 2.12:

A model of the atom shown in Preston's 1906 textbook of heat.

Clifton's views on the experimental facilities in Quay Street were still being talked about when Arthur Schuster arrived in 1871, five years after Clifton had left. 'The only research he could carry out in the so-called laboratory' reported Schuster, 'was to determine the kind of vehicle that was passing in the street below, by observing the nature of the seismic disturbance'. Motor vehicles did not arrive on the scene until 1896, so the traffic was horse drawn. The large four wheeled cabs were called 'growlers' on account of the noise they made. It was an unusually pithy observation by Schuster, who otherwise always comes across as a gentle, compassionate person. In addition, Schuster had an advantage denied Clifton who was compelled to work in an impossibly cramped 'laboratory' in Cobden's house. The move to the spacious buildings on Oxford Road did not take place till 1873.

Clifton also wrote a maths paper with the title *Note on Prof de Morgan's paper entitled 'On the early history of the signs + and −'*, published in the 1866 Cambridge Philosophical Transactions. Four years earlier, he had published a joint paper with the chemistry professor Roscoe: *Effect of increased temperature upon the nature of the light emitted by the vapours of certain metals or metallic compounds*. In this paper, they argued that the flame spectra were due to oxides and not the elements themselves. This also appeared in the Manchester Lit. and Phil. *Proceedings*. His total publication record during his five years in Manchester was almost an order of magnitude greater than his one and only paper, published during his 50 years as professor at Oxford.

He carried a full teaching load and it is for teaching rather than research that he is best remembered. During the first few decades after its founding, the Calendar of Owens College carried details of the curriculum and the edition for the session 1863-4 contains the following entry:

'NATURAL PHILOSOPHY
Professor R B Clifton, MA, FRAS, FCPS
Experimental course
Mechanics – Tuesday and Friday from 12.30 to 1.20 p.m.
In this class the principal proposition in Statics, Dynamics,
Hydrostatics and Hydrodynamics will be considered and explained
by the aid of experiment. The course will include the portions of
these subjects required of the candidates for Matriculation.
Fee, £3. 3s.
Physics – Monday and Wednesday from 2 to 3 p.m.
In this class the Professor will explain, and illustrate by
experiment, the phenomenon observed in the following branches of
Physics: Heat, Acoustics, Optics (Geometrical and Physical).
Fee, £1. 11s 6d.'

Similar courses were held for an hour on Wednesday evenings and the students were told what was expected of them, in an exhortation of mind-blunting grammatical clumsiness:

'It is to be clearly understood that to no one joining any of these classes can any good whatever come from simply attending, however well and regularly; at the class hour. Much time must be given besides to prepare for the stated meetings by thinking out, and working diligently at the exercises set in illustration of the matters handled in the class.'

Some of the exam questions associated with this course, published in full in the annual College Calendar (without answers) were quite tough and some have become impenetrable with the passage of time.

'Q2. Enunciate the *Parallelogram of Forces.* Explain the mechanics involved in raising and flying a kite. When the kite is flying, what will be the effect of suddenly pulling the string? Give reasons for your answer.'

'Q6. Describe the steelyard, and show it must be graduated.'

'Q13. Describe the construction and action of an overshot waterwheel; and point out the main advantages and disadvantages of this form of waterwheel.'

Question 13 was a gift for anyone who had read Smeaton's 1759 paper on overshot (and undershot) water wheels, as we saw in Chapter 1.

One of the most lasting achievements of Clifton, just before he left, was to provide quantitative proof that the existing college building, the converted house of Richard Cobden, was no longer large enough to contain the classes safely, thereby helping the case for the eventual new college. In the session 1864/5, he was lecturing in physics to a day-class of 47 out of a total of 127 students attending the college. Even today, some Universities have smaller classes in physics. Evening classes were attended by a total of 439 students, probably not all at once. Clifton concentrated on the day class and with the help of the chemist Roscoe, made some measurements. 49 people, including the professor and the demonstrator, Francis Kingdon, were gathered into the physics lecture room, which had a volume of 5,200 cubic feet. Assuming a ceiling height of 12 ft, (it was gentleman Richard Cobden's former house, after all), the average floor space per person of $0.82\,m^2$ exceeded the modern standard of 0.6, so there was room enough. With 100 cu ft of air available per person per hour in the room, and the gas lighting burning twice as much oxygen as the people, Clifton concluded that only extreme ventilation could keep the class alive. With the windows wide open, even in winter, and most of the class wearing Crimean war army surplus great coats (student J J Thomson had one six years later),

Roscoe measured the temperature and CO_2 levels at the beginning and end of the lecture.

The students (and the staff) were greeted at the start of the lecture with a room at an ambient temperature of 5°C and a CO_2 level of 0.03%. By the end of the lecture, the temperature was 18°C and the CO_2 level was 0.29% at ground level and 0.35% at the ceiling. Clifton estimated that it would be 0.32–0.33% at head height and went on to calculate that the five gas lights would contribute 1% to the CO_2 levels every hour, adding to the 0.5% exhaled by the occupants. Hence, since it had only reached 0.35%, he was able to conclude that the extreme ventilation was effective, albeit at the cost of temperature comfort. These figures indicate that the ceiling height was more likely to be 8 ft than 12 and J J Thomson's account of Richard Cobden's house [97] gives a clue:

'The stable itself was converted into the Lecture Room, and the hayloft above it into the Drawing Office, which had to be reached by an outside uncovered wooden staircase.'

It is currently held that when the concentration of carbon dioxide is 1%, as in a full auditorium with poor ventilation, some occupants are likely to feel drowsy. Above 2%, most occupants will be affected and above 3%, breathing rates will be double what is normal. By way of interest, until about the year 1800, the atmosphere contained about 0.03% CO_2, compared to its current level of 0.04%.

Clifton was in Manchester for no more than five years before he was appointed to the chair of experimental philosophy at Oxford in 1865. 'Too soon Owens had to part with him to Oxford', wrote Roscoe [8] in 1906, 'a University which he still adorns'. It is possible that Roscoe was being politically smart or facetious because Clifton is remembered in Oxford as the man who stifled research during his long tenure. When he finally retired at the age of 80, research equipment delivered at the time of his appointment 50 years previously was found still unpacked in a cupboard. Rudolf Peierls told me, quoting Pauli, 'He published only one paper, and it wasn't even wrong.' Not only was it not wrong, but Clifton's paper of 1877 on the source of the electromotive force of a dry cell, brought the wrath of William Ayrton and John Perry (see Chapter 3 for more on Ayrton and Perry) upon his head, since they had already published the same thing the previous year. He never dared publish again.

Graeme Gooday, in his article [100] *Robert Bellamy Clifton and the 'Depressing Inheritance' of the Clarendon Laboratory, 1877–1919* goes some way to ameliorate this view by suggesting that Clifton 'generally subordinated research to the higher imperative of training students to the most rigorous standards of experimental physics'. It is of course, not totally fair to compare the current top universities in the world, which are primarily research driven and find that quality teaching inexorably follows quality research, with Oxford 100 years ago. But this overly generous view of Clifton overlooks the fact that if there is no research, there will ultimately be nothing to teach, except in the style of chapters of Euclid's treatise. This might have been his intention since he was known to hoard old Euclids in his personal library which he named his 'Folly'.

Clifton himself stated the position in Oxford succinctly in 1910, more than four decades after leaving Manchester:

> 'All the available space in the Laboratory is now devoted to the instruction of students preparing for the examination in the School of Natural Science, and it will in future be impossible to offer facilities to advanced students wishing to engage in Research.'

a sentiment from Oxford through the mouthpiece of Clifton, uttered in the same year that Rutherford's *undergraduate* student Marsden, was helping to discover the nucleus in the research laboratory space that Rutherford inherited from Schuster (see Chapter 5 in Vol. 3 to come).

It took Clifton 46 diligent years at Oxford to get to the situation where he made this remark, surely not without deliberation and choice. Even if, as it has been argued [100], he was thwarted at Oxford by academic politics and funding problems, it was always open to him to use the ploy that got him his chair at Oxford in the first place – get some influential scientists to write letters to the University governors. Indeed, Gabriel Stokes, William Thomson, James Joule, Henry Roscoe, Wilhelm Bunsen and Gustav Kirchhoff had all written glowing references on his appointment [101] and hoped that Oxford would offer greater scope for research than had been possible at Manchester. Had Clifton known of the content, all he had to do was ask them to write again to the University governors, demanding to know why his research was being obstructed and the problem would surely have been solved. In the event, the prognosis of some of the most eminent scientists in the world concerning the relative research opportunities in Manchester and Oxford could not have been more wrong. Yet Clifton himself determined the opportunities. Casting unworthy thoughts aside that Roscoe might have asked his friends, Stokes, Bunsen and Kirchhoff, to help Manchester be rid of a mill-stone, whilst Joule did the same with his close friend William Thomson, it is possible that all of them genuinely thought that Clifton, hired by Manchester at the age of 24 still had considerable potential at the age of 29 when he left for Oxford.

It is ironic that Clifton's next but one successor Balfour Stewart, who followed the abysmal and stagnant four years of William Jack, created the research laboratory space in which even a Clifton must have thrived.

Despite all these negative aspects, Clifton was well liked. A Manchester newspaper, reporting the appointment of his successor William Jack said of the incoming and outgoing professors:

'There is in him little of that easy grace which marked Mr Clifton, making him most interesting and dignified when lecturing from a seat on the table, or when, having thrown away his gown, he worked with a vigour quite astounding to the audience at some laborious experiment.'

O F Brown, a demonstrator in the Clarendon Laboratory, 1910–13, generously reminisced [102] on the Clifton he knew at that time:

'... a delightful old gentleman and kindness itself. But for him physics had reached its end in the 1890s'

According to *Who's Who*, Clifton's recreation was 'Work'. He would begin his working 'day' when his household went to bed and go right on until about 8 am. He would then sleep for a couple of hours or so, before showing up in the department around 11 am. After dinner, he would rest in his chair until it was bedtime for the others.

It is a sobering thought for Manchester that if Clifton had not gone to Oxford in 1865/66 but had remained at Owens for the next 50 years, there would have been no Schuster to fill the Langworthy Chair and to establish one of the largest physics laboratories in the world. There would have been no Rutherford at Manchester to discover the nucleus, no Bohr to write his trilogy, no Moseley to set the Periodic table in order and no Bragg to make X-ray crystallography into a routine science capable of unraveling the structure of DNA. There would probably have been no reason for Blackett to come to such an ordinary place as Manchester and whence there would have been no discovery of strange particles, which heralded the unexpected 3rd quark and since Blackett hired Bernard Lovell, no Jodrell Bank either. Gooday [100] decided 'not to consider what research Clifton might have done at Oxford, but rather what he did do.' All the evidence suggests that the research that Clifton might have done can be gleaned from his few presentations to the Manchester Lit. and Phil. From the Manchester point of view, it is a joy and honour to realise what discoveries were made without him and a shuddering relief to know what would not have been done, with him.

Manchester records Clifton as having resigned in 1866. Oxford records his appointment there as 1865. Physically, he was still in Manchester on the 14th of November 1865 when he presented a paper to the Lit. and Phil.. Overlapping appointments and double salary were common at the time.

1866: William Jack

Figure 2.13:

William Jack, Professor of Natural Philosophy at Owens College, 1866–1870.

Colour processing © 2017 Robin Marshall.

Upon Clifton's departure from Manchester, Scotsman William Jack (see Figure 2.13), a mathematician who became a lifelong friend of Roscoe, was appointed to the vacant chair of Natural Philosophy. Given the exertions that went into extracting the teaching of physics from the grip of Sandeman, it seems curious to have reverted to a mathematician to fill the post. In his brief stay in Manchester, Jack did not publish a single paper on physics. One could misquote Pauli and say: 'He did not publish a single paper and that was not right.' I have scoured the pages of the *Transactions and Memoirs of the Manchester Lit. and Phil.* during his tenure but apart from a year that he spent on the Society's Council, there is not a paper nor summary of any talk by him. The Lit. and Phil. was the hub of Manchester science discussions and he seems to have played little if any part.

After only four years in post, he resigned to become editor of the Glasgow Herald in 1870 in order to pursue his desire for a more public career. The fellowship of St. Peter's (Peterhouse) that he held whilst in Manchester ended the following year although the academic connection was not totally severed, since he was awarded an honorary degree of LL.D. by Glasgow whilst still an editor. He was also invited onto the board of governors of the extended Owens College when it moved to its new site in Oxford Road in 1873. After six years as editor, he was briefly a partner with the publishing firm MacMillan in London before accepting a chair of mathematics at Glasgow University in 1879. He was appointed to this chair partly on the recommendation of William Thomson who had originally recommended his admission to St Peter's in 1855.

For his 1879 inaugural lecture in Glasgow, [103] he referred to his former employers, Manchester, in the same breath as the subject of his lecture – Galileo:

'It was my fortune also to be charged for a short time with the teaching of Physics in Manchester, where we have seen a University grow up in our own generation substantially after our type, and which promises one day to rival the fame it has taken us centuries to accumulate. I may perhaps be excused, then, for selecting as my subject the work of the great Italian, who did more perhaps than any other man, with the doubtful exception of our own Newton, to apply mathematics to the most splendid task which it was ever proposed to it, who victoriously solved the fundamental problems of mechanics and Astronomy, and laid deep and broad the foundations of Physical Science.'

Thus did William Jack, whilst giving his *Alma mater* a poke in the eye, demonstrate his view of the relationship between mathematics and physics. Not having done any research in physics, nor indicated any inclination to do any, it would have been interesting to hear a lecture by him on the origins of the theorems of geometry, which would not be there but for the measurements of geometers. The closest he got in this lecture to accepting this fact was to say:

'Measurement even in its simplest form, mathematics, or, if you choose, arithmetic, lies at the root of all our knowledge of nature.'

The discoverers of the structure of DNA or the Higgs boson surely never regarded their efforts as simple arithmetic. The early warning in his lecture that it was doubtful if Newton were superior to Galileo, (not that it really matters), was followed by undiluted sycophancy. There was no room in the 33 pages to credit 'The Law of Falling Bodies' and similar works to the 'Oxford Calculators' at Merton College from about 1330 onwards. One can ponder if Jack had ever heard of Thomas Bradwardine, William Heytesbury, Richard Swineshead and John Dumbleton.

The bulk of Jack's career was subsequently spent as Professor of Mathematics at Glasgow University, a post he held for 30 years, eventually retiring in 1909 at the age of 75 because his strength 'was beginning to fail'. There was no compulsory retirement at 67, 65 or 60 in those days. He enjoyed 15 years of retirement.

William Jack had entered St Peter's College/Peterhouse Cambridge in 1855. Already 21, he was older than his student contemporaries, one of whom was Adolphus (later Sir) William Ward. Ward was appointed to the chairs of History and English Language and Literature at Manchester in 1870, the same year that Jack was appointed to the chair of Natural Philosophy. Ward himself remained at Owens throughout its move to the Oxford Road site and its subsequent achievement of University status, eventually becoming Principal of the University in 1889. In 1924, he wrote an obituary about Jack in the Peterhouse sexcentenary magazine *The Sex* [104].

The obituary is both generous and honest. Ward wrote that William Jack obtained a brilliant degree in 1859, being Fourth Wrangler and First Smith's Prizeman. These achievements, Ward said, put his own competition for a vacant College fellowship out of the question and Jack was duly elected. Once at Manchester, Ward continues, Jack's combined sagacity and courage took him to a leading position in the Senate of the College where he put his natural gift of speech and powers as a debater to good use.

Having deferred to a better mathematician in his obituary, talking about Jack's academic achievements, Ward was candid about his term as editor:

> 'His editorship had soon fallen on troublous times ... and though his success as an editor may not have been comparable to that of some others of his generation, the standard of daily journalism was never lowered in his hands.'

He was always a member of the several delegations that went to Westminster to argue for public funds for the extension of the Owens College and for its eventual granting of a University Charter. He was invited to the Jubilee Celebrations in 1901 [5] as a distinguished guest.

One can be as equally candid about Jack's term as a physics professor as Ward was about his editorship of the Glasgow Herald. The fact remains that the department of Natural Philosophy at Owens stagnated under Jack's leadership, whilst all the time, he was orating impressively at Senate and unproductively at Westminster. Manchester physics did not thrive and barely survived during the 13 years of the combined stewardship of Scotsmen Sandeman and Jack.

He became a member of the Manchester Lit. and Phil. immediately on appointment and the initial portents were good with a paper [105] on the non trivial mathematics of the angular deflection of a galvanometer needle as a function of the passing current, the radius of the coil wire, the length of the needle and various other environmental factors. He had clearly been in regular discussions with Joule, who sought the most precise instrument possible. A paper by Joule on the subject immediately followed Jack's in the proceedings and this, together with further exchanges, led to two further papers by Jack on the subject [105]. The final paper alas, written in the 3rd person, possibly by the society's secretary, started with the ominous paragraph:

> 'The author, referring to his communication entitled "Further Remarks on the Galvanometer," printed in the last number of the Societys Proceedings, points out that a serious misprint had occurred on page 159, where in the third line $n + i$ should be n and i.'

The rest of the paper pointed out and emphasised how important the formula was, despite the adjoint error. In 1868, he presented a review of Peter Guthrie Tait's work on thermodynamics to the society and that was the end of his publication record.

Mention of Tait provides an opportunity to mention the Reverend Cosmo Reid Gordon, one time rector at Chetwynd in Shropshire, who applied for the Chair of Natural Philosophy at Edinburgh University in 1860 along with James Clerk Maxwell (Aberdeen), E J Routh (Cambridge), Tait himself (who got the job) and others. Gordon submitted his application from Manchester. Tait's biographer Cargill Gilston Knott remarked [106] 'There is no difficulty now about placing these men in their appropriate niches in the Temple of Fame', and proceeded to explain why Tait (and Maxwell) were superior to Fuller and Routh, whilst consigning Gordon to the indignity of not even being mentioned. At the time, the Edinburgh *Courant* newspaper agreed with Tait being preferred over Maxwell on the grounds of the latter's 'deficiency of the power of …

oral exposition'. But of Cosmo Reid Gordon, there is no mention nor trace. Whatever physics reasons he had for applying for the distinguished chair from a putative base in Manchester remain unearthed. But his name will for ever be associated, albeit tenuously and in vain, with Manchester physics.

Meanwhile Jack, during his tenure from 1866 to 1870, was completely out-published by James Joule, James Benjamin Dancer, James Nasmyth and William Fairbairn, in the society's journals. And lest it be claimed that teaching made research difficult, his contemporary Osborne Reynolds out-published them all.

All this was about to change, and it would need another Scotsman, Balfour Stewart, essentially an astronomer, to make the change in what was about to become an actual department rather than a sphere of influence of a professor. The story of the progression of Manchester physics under Balfour Stewart will be continued in the first chapter of the next volume. But first, we may not let the above mentioned Osborne Reynolds go further unmentioned.

1868: Osborne Reynolds

Osborne Reynolds was hand-picked and appointed to the post of 'Professor of Engineering' in 1868. He appears in the Senate group in 1872, (see Chapter 3 in Volume 2) at a time when he was teaching J J Thomson. The chair was funded by local industrialists and there being no department of engineering as such at the time, Reynolds was closely associated with the department of natural philosophy. The old Quay Street building did not support the concept of separate departments and all disciplines, arts and sciences were compelled to rub shoulders.

Osborne Reynolds was born in Belfast on the 23rd of August 1842, the son of the Rev Osborne Reynolds who had been thirteenth Wrangler in 1837 and then a fellow of Queens' College, Cambridge. Son Osborne was seventh Wrangler. He is best remembered for the number which bears his name, known to all physicists and engineers, defining the point at which a fluid flowing smoothly through a tube becomes turbulent.

He was a regular attender at the Lit. and Phil. meetings throughout his career, making prolific and eclectic contributions to their proceedings. Just as the so-called chemist Dalton was a physicist as well, so was the so-called engineer Reynolds also a physicist.

A paper he presented [107] to the Lit. and Phil. on the 20th of February 1872 demonstrates his refusal to allow what others might regard as common sense getting in the way of presenting a novel idea. He set up to demonstrate that he was able to induce a brass ball in an evacuated vessel to emit what he claimed to resemble the corona of the sun, which at the time was not understood and even held by some to be an optical effect of the earth's atmosphere. This would not be the last time that Reynolds sought to bring astronomical and cosmological phenomena onto the laboratory bench.

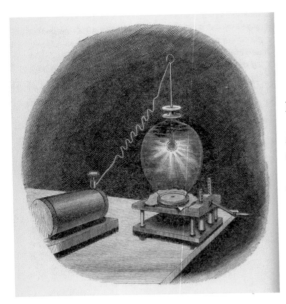

Figure 2.14:

Reynolds' demonstration of how the sun's corona might be produced.

Image taken from reference [107].

Here, he applied a voltage from a battery cell or a machine to a brass sphere in a partially evacuated vessel (see Figure 2.14) and observed how the corona changed their appearance as he let air in, a flickering array of twining and untwining serpents. Remarking that he only used one pole and let the negative electricity find its own path, he mused on what became of it. The experiment may have had one profound effect. J J Thomson became a student in the college at that time and Reynolds would surely have shown off this apparatus to his class. Thomson went off, with whatever impression it made on him and a quarter of a century later showed that the negative electricity consisted of electrons. If Reynolds had focussed on the nature and production mechanism of the little corona he produced, the discovery of the electron might have been his. But apart from his lifelong devotion to fluid dynamics, he had a tendency to flit superficially from one topic to another.

He also devised an ingenious experiment to test the behaviour of the radiometer or 'light mill', a device invented by William Crookes in the

152

1870s. Crookes had devised the radiometer whilst carrying out painstaking measurements, weighing samples of thallium, an element he had recently discovered, on a delicate balance. By careful work, he realised that certain discrepancies he was observing were caused by light falling on the balance and affecting the reading. Now according to Maxwell's treatise on electricity and magnetism, published in 1873, light ought to exert a pressure when it falls on a surface. It was natural for Crookes to assume that the effect he was seeing was a material manifestation and proof of Maxwell's theory.

Crookes therefore built a small device with four vanes of mica fastened on the ends of a cross which was then mounted on a pivot supported on a glass cup in such a way that the cross and the vanes could rotate in a horizontal plane. This structure was enclosed in a glass vessel which was evacuated to the lowest pressure possible. Reynolds had discussed the problem with Schuster and they had done some experiments in the physics department about which Reynolds wrote letters to *Nature* suggesting Crookes was wrong. Crookes was having none of it and a lengthy exchange of correspondence via *Nature* ensued. Schuster then inadvisedly entered the fray to defend Reynolds against Crookes' robust criticism by also writing to *Nature*. Crookes demolished them both at a stroke by saying that whilst reading the distinguished Professor Schuster's words, all he could hear was Reynolds' voice – which was wrong.

As a lecturer, Reynolds was quite undisciplined. J J Thomson, whilst praising the quality and stimulating effect of a Reynolds lecture, gives a vivid picture of what it was like to attend one:

> 'Ten minutes into the lecture, with no sign of Reynolds, the students would go and get the janitor to find him. He would enter the room in a swirl, still pulling on his gown, and turn at random to a page in one of Rankine's textbooks. His first act would be to declare that the formula he had just fallen upon was wrong from whence he proceeded to the blackboard to prove it. His style is not unfamiliar, even today, talking to himself at the blackboard with his back to the students. After a few seconds he would wipe the board clean and say it was wrong. The whole procedure was recommenced and would be repeated throughout the lecture until finally a formula would survive unscathed. At this point he would declare that Rankine was right after all.'

Thomson went on to say that although Reynolds' lectures did not usually increase the students' knowledge of facts, they were interesting because they showed an acute mind grappling with new problems.

On the 10th of June 1902, Reynolds presented the Rede Lecture at Cambridge entitled: *On an Inversion of Ideas as to the Structure of the Universe.* The subject had become his driving force during the previous 20 years. He published the lecture in the form of a book [108], albeit a slim one. The book itself gives little insight into how Reynolds fell on this idea nor how he deduced his ideas on what the universe, and in particular the aether was made of. The results are presented as fact. The lecture and book received lengthy reviews on account of Reynolds' reputation.

The essence of his theory was that the universe behaved like a continuum of closely packed little spheres. When subjected to a compression, a medium of closely packed spheres can only expand. There are two manifestations of this phenomenon that are known in everyday life. When a foot is pressed into wet sand, the sand will expand and form a dry border to the foot. More striking, a vacuum pack of coffee beans or grind feels hard when squeezed because compression in resisted. Reynolds accompanied his lecture with a demonstration of the phenomenon which he called 'dilatancy' (see Figure 2.15).

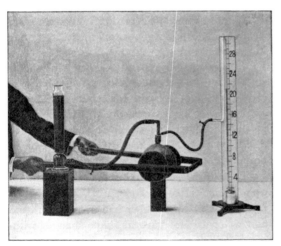

Figure 2.15:

Reynolds' demonstration of how the Universe works. If the Universe is squeezed, its volume expands, as shown by the barometer.

Reynolds had used the classic ruse of a magician, distracting the audience with something that is true but eye-catching and then carrying off the deception on the back of the distraction. But to be equally honest, Reynolds' model had some merits even by modern standards. The aether was essentially the incompressible medium and light propagated by means of dislocations in the otherwise closely packed structure. The size of the closely packed particles was precisely 5.534^{-18} cm, smaller than the smallest resolvable size in modern particle physics experiments.

The oil portrait of Reynolds copied in Figure 2.16 shows him holding the 'whole Universe' in his hands.

Figure 2.16:

Osborne Reynolds in 1902. He is holding a representation of his model of the Universe.

This image has been constructed by forming a stack of individual images in the public domain, each component of the stack inferior to the final result.

This image is declared by Wikimedia Commons to be in the public domain.

The slim, qualitative booklet of his Rede lecture was followed by a more substantial mathematical treatment entitled *The Sub-mechanics of the Universe* [109], whose mathematics put Maxwell's Treatise to shame. Sharing the Lords Rayleigh and Kelvin's view that physics in 1900 was essentially sorted out and just a few frayed edges needed tidying up, Reynolds had no idea how un-prophetic his opening sentence would soon turn out to be:

> 'By this research it is shown that there is one, and only one, conceivable purely mechanical system capable of accounting for all the physical evidence, as we know it, in the Universe.'

Even Schuster knew in 1900 that the energy in the Sun could not be explained by any known phenomena.

Although he emitted occasional flashes of eccentricity, Reynolds was an accomplished and professional engineer and physicist, indeed a true natural philosopher. In 1892, the Lit. and Phil. published a memoir by him on the life and work of James Prescott Joule. As a reasoned argument and presentation of the development of the physics of heat and energy during the 19th century, the memoir of 196 pages is both a book and, like Cardwell's biography [15] a *tour de force*. Moreover, it contains a print of the engraving of the 1863 oil painting of Joule, which was destroyed during an air raid in 1940, from which I have re-created a new (acrylic) portrait.

Physicist-engineer Osborne Reynolds started his career in Manchester in 1862, two years after Archibald Sandeman had responsibility for physics taken from him, and he ended it in 1905, two years before the mantel of physics was given to Ernest Rutherford.

1853–1856: The Crimea War Veteran

The Manchester Physics Department photographic archives contain a glass plate, reproduced in Figure 2.17 on the next page. It is easy to establish that the subject is a Crimea war veteran from the medals, but despite the physics legend, the connexion to the department cannot be established. The three medals can be identified by magnification.

The medal on the left is the 'Crimea War Medal' with one clasp. Even with image processing and enhancement, it is not possible to discern to which campaign the clasp relates, except that it almost certainly is not Alma nor Azoff, on account of the length of the text, even though it is blurred. The clasp is therefore of Inkerman, Balaklava or Sebastopol, any of which would indicate that he landed in the Crimea before the 9th of September 1855.

The central medal was awarded for 'Long Service and Good Conduct (Army)', 18 years being the qualifying period. The medal on the right, tilted, thus making identification more difficult, appears to be the 'Turkish Crimea War Medal' which was awarded by Sultan Abdülmecid I of the Ottoman Empire to all allied military personnel involved in the war. There were three versions of this medal, the British, the French and the Sardinian issue. To further complicate the story, most medals struck for the British soldiers were lost when the ship carrying them was sunk. Assuming that the veteran here is British, he would probably have received whatever was left over from the batch of French or Sardinian medals which were of poor quality, suspended from a narrow ribbon by a circular steel ring and prone to rusting. British Officers frequently had superior copies made in 925 silver with suspenders that were plain, in comparison with the other medals worn by our hero here which are more ornately suspended. As well as replacing the medal with a better silver version, the narrow ribbon was often replaced with one of standard British medal width so that it hung properly. In this photograph, the Turkish medal is indeed suspended from a British standard width ribbon, but by having only a simple suspender, without an additional ring, the medal then hung tilted.

PHYSICS DEPT.
MANCHESTER UNIVERSITY

Figure 2.17:

Someone with a Manchester Physics and Crimea War connexion, wearing campaign and service medals.

Many thousands, perhaps tens or even hundreds of thousands of British soldiers could have worn all these three medals together and so they do not help to identify the veteran. We can possibly conclude that he was not an officer, from his artisan clothing, nor was he likely to have been a physics professor, since he served at least 18 years with irreproachable character and conduct in the British Army.

Being a Crimea war veteran, we could say that he might have been about 25 during the campaign. If we guess at the age of 65–75 in the photograph here (people looked older than their clock age then than they do now), then we have a date of 1895–1905 plus or minus a decade, for when this photograph was taken.

A diligent search has revealed no provenance for this image. The only relevant information elucidated, comes from the Illustrated London News (ILN) archives, which reveal that in 1896, a Captain Hobbs organised the third of a series of Crimea veterans reunions in Manchester. 220 veterans paraded along the esplanade in front of the Manchester Infirmary in Piccadilly to the rapture of applauding crowds, and were then treated to a dinner in the Town Hall. On the first two of these occasions, a telegram of greetings had been sent to the Queen but in 1896, it was deemed not propitious to do so, since she was entertaining the Czar of Russia to whom she was closely related, but politically estranged on account of the war decades ago.

Czar Nicholas was a physical clone of King George V, up to dietary differences, a situation possibly encouraged by the relentless inbreeding among European royalty at the time. Nicholas, his wife Alexandra and George V were all cousins, not to mention Kaiser Wilhelm, who differentiated at least by using a unique barber. Nicholas even had what history records coyly as a 'flirtation' with another English first cousin of his, Victoria, younger sister of George V. The cognatial attraction amongst royalty during this period seems stronger than that between quarks, although this may be the only connexion to physics.

But clearly, none of this has anything to do with Manchester physics, at least superficially, apart from the veteran thus labelled. By placing him on the record, one day his identity might be revealed.

Bibliography

[1] Joseph Thompson. *The Owens College; its Foundation and Growth.* Manchester: J E Cornish 1886.

[2] Joseph Thompson. *Lancashire Independent College, 1843–1893.* Manchester: J E Cornish 1893.

[3] William Arthur Shaw. *Manchester Old and New.* Cassell and Company Ltd., London 1894.

[4] P J Hartog (Ed.) *The Owens College, Manchester: (Founded 1851) A brief history of the college and description of its various departments.* J E Cornish, Manchester 1901.

[5] Josephine Laidler. *Record of the Jubilee Celebrations at Owens College Manchester.* Issued at the request of the Council by the Committee of the Owens College Union Magazine. Sherratt & Hughes, Manchester 1902.

[6] *The Physical Laboratories of the University of Manchester. A Record of 25 Years' Work.* Manchester at the University Press, Manchester 1906.

[7] Sir Arthur Schuster. *Biographical Fragments.* MacMillan and Co. Ltd. London 1932.

[8] H E Roscoe. *The Life and Experiences of Sir Henry Enfield Roscoe. Written by himself.* MacMillan, London and New York 1906.

[9] *The Court Leet Records of the Manor of Manchester from the Year 1552 to the Year 1686 and from the Year 1731 to the Year 1846.* Printed under the superintendence of a committee appointed by the Municipal Council of the City of Manchester, from the original minute books in their possession. Volumes I to XII, Henry Blacklock & Co., Manchester 1884.

[10] Henry A Bright. *A Historical Sketch of Warrington Academy.* Liverpool 1859.

[11] William E A Axon. *The Annals of Manchester. A chronological record from the earliest times to the end of 1885.* Manchester: John Heywood 1886.

[12] Henry Buckley Charlton. *Portrait of a University: 1851–1951: To celebrate the centenary of Manchester University.* University Press. 1951.

[13] D S L Cardwell (Ed). *Artisan to graduate. Essays to commemorate the foundation in 1824 of the Manchester Mechanics' Institution.* Manchester University Press 1974.

[14] Richard L Hills (Ed). *The Development of Science and Technology in Nineteenth-Century Britain. The Importance of Manchester.* Ashgate Publishing Limited. Aldershot 2003.

[15] Donald S L Cardwell. *James Joule. A Biography.* Manchester University Press, Manchester 1989.

[16] Robin Marshall *Physicists at War.* Champagne Cat. 2018.

[17] Robin Marshall. *The Colouring of Archibald Sandeman.* Journal of the Perthshire Society of Natural Science. Vol **XX**. pp5–14. November 2014.

[18] Osborne Reynolds. *Memoir of James Prescott Joule.* Memoirs and Proc of the Manchester Literary & Philosophical Society. Series 4. **6**. 1892.

[19] J T Fowler. *Durham University; earlier foundations and present colleges.* F E Robinson & Co, London 1904.

[20] John Rushworth. *Historical collections containing the principal matters which happened from the dissolution of the Parliament on the 10th of March $162\frac{8}{9}$ until the summoning of another Parliament which met at Westminster, April 13, 1640.* Wright, Chiswell. London 1680.

[21] George W Johnson (ed). *The Fairfax Correspondence. Memoirs of the Reign of Charles The First.* Richard Bentley, Publisher in Ordinary to her Majesty. London 1848.

[22] Samuel Hibbert, John Palmer, William Robert Whatton, J. Greswell. *History of the foundations in Manchester of Christ's College, Chetham's Hospital and The Free Grammar School. Volume 1.* Thomas Agnew and Joseph Zanetti. Manchester. 1830.

[23] John Palmer. *The History of the Siege of Manchester.* Longman & Co., London 1822.

[24] Henry Taylor. *Old Halls in Lancashire and Cheshire..* J. E. Cornish, Manchester 1884.

[25] John Towill Rutt (ed). *Diary of Thomas Burton, Esq. Member in the Parliaments of Oliver and Richard Cromwell from 1656 to 1659.* Vol II. Henry Colburn, London 1828. pp531–543.

[26] John Richard Green. *A Short History of the English People.* MacMillan and Co. Ltd., London 1888.

[27] Edmund Calamy. *An Account of the Ministers, Lecturers, Masters and Fellows of Colleges and Schoolmasters, who were Ejected or Silenced after the restoration in 1660.* London 1713.

[28] Henry Kirke (ed). *Extracts from the Diary and Autobiography of the Rev James Clegg (Non-conformist Minister and Doctor of Medicine.* Buxton, C F Wardley. London Sampson Low, Marston. 1899.

[29] Dennis Porter. *A catalogue of manuscripts in Harris Manchester College, Oxford.* Harris Manchester College 1998, updated 2009.

[30] Irene Parker. *Dissenting Academies in England.* Cambridge University Press. 1914.

[31] J E Odgers (ed). *Manchester College, Oxford. Proceedings and addresses on the occasion of the opening of the college buildings and dedication of the chase. October 18-19 1893.* Longmans, Green 1894.

[32] V D Davis. *A History of Manchester College from its Foundation in Manchester to its Establishment in Oxford.* George Allen. London. 1932.

[33] J L Heilbron. *Electricity in the 17th and 18th Centuries: A Study of early Modern Physics.* University of California Press. Berkeley, Los Angeles and London. 1979.

[34] John Farey. *General View of the Agriculture and Minerals of Derbyshire.* Vol 1. Drawn up for the consideration of The Board of Agriculture. London 1811.

[35] *The Journal of the Rev John Wesley. Vol II. November 25 1746 to May 5 1760.* London. J Kershaw. 1827.

[36] John Neale. *Directions for Gentlemen who have Electrical Machines.* London 1747.
Seventeen electrical experiments for a gentleman to perform with plants and animals. Gentleman's Magazine, xvii March 1747.

[37] M l'Abbé Nollet. *Essai sur l'Électricité des Corps'.* Paris 1750.

[38] George Adams. *Lectures on Natural and Experimental philosophy.* 1806. Whitehall and Philadelphia 1807.

[39] Albert Edward Musson and Eric Robinson. *Science and Technology in the Industrial Revolution.* Manchester University Press. 1969.

[40] A Walker. *Analysis of a Course of Lectures on Natural and Experimental Philosophy.* 6th ed. Printed and published by the author. Undated. (~1780).

[41] J Gough. *A description of a property of caoutchouc or Indian rubber; with some reflections on the cause of the elasticity of this substance.* Memoirs of the Literary and Philosophical Society of Manchester. Series 2. **1**. pp288–295.

[42] *Memoirs of the Literary and Philosophical Society of Manchester.* Vol **2**. Warrington 1785.

[43] Rudolph Ackerman. *The Microcosm of London. Volume III.* Methuen & Co. London 1904.

[44] Thomas Leybourn of the Royal Military College. *New Series of the Mathematical Repository.* W Glendinning. London 1806.

[45] Peter Barlow. *A New Mathematical and Philosophical Dictionary; etc.* London 1814.

[46] Bonamy Dobrèe ed. *The Letters of King George III.* Cassell, London. 1935. p212.

[47] *The Monthly Magazine and British Register for 1797.* Vol. **4**. London. 1798.

[48] John Seed. *Manchester College, York: An Early Nineteenth Century Dissenting Academy.* Journal of Educational Administration and History, **14**. 2006. pp9–17.

[49] William Charles Henry. *The Life and Scientific Researches of John Dalton.* Cavendish Society, London. 1854.

[50] John Dalton. *Memoirs of the Literary and Philosophical Society of Manchester.* Vol. **5** Part 2. 1802. pp 535-602.

[51] J L Gay-Lussac. *Recherches sur la dilatation des gaz et des vapeurs.* Annales de chimie **43**. 1802. pp137-175.

[52] John Dalton. *A new system of chemical philosophy. Part 1.* Manchester 1808.

[53] Silvanus Phillips Thompson '*Lectures on the Electromagnet.*' W J Johnston & Co. Ltd. February 1891.

[54] J P Joule. *A short account of the life and writings of the late Mr William Sturgeon.* Memoirs of the Literary and Philosophical Society of Manchester. 2nd series **14** 1857. pp53–83.

[55] W Sturgeon. *Improved electromagnetic apparatus.* Transactions of the Society, Instituted at London, for the Encouragement of Arts, Manufactures, and Commerce, **43**. 1824, pp37–52 and Plates 3 and 4 in the same volume.

[56] Wm Robert Whatton. *An Address to the Governors of the Royal Institution of Manchester, containing proposals for altering and extending the present plan of the Institution and for giving it the power and efficient form of an University.* Henry Smith, Manchester 1829.

[57] Wm Fairbairn. *Observations on improvements of the town of Manchester, particularly as regards the importance of blending in those improvements, the chaste and beautiful, with the ornamental and useful.* Robert Robinson, Manchester 1836.

[58] Harry Longueville Jones. *Plan of a University for the town of Manchester.* Pamphlet of a paper read before the Manchester Statistical Society 1836.

[59] Joseph Priestley. *A Familiar Introduction to the Theory and Practice of Perspective.* J Johnson and J Payne. London. 1770.

[60] James Nasmyth and James Carpenter. *The Moon considered as a Planet, a World, and a Satellite.* John Murray, London 1885.

[61] James Nasmyth. *On the planet Mars.* Memoirs of the Literary and Philosophical Society of Manchester. Third Series. Volume **2**. 1865. pp303–305.

[62] James Nasmyth. *James Nasmyth Engineer. An autobiography.* John Murray, London. 1883.

[63] C F Bartholomew. *The discovery of solar granulation.* Q. Journal Royal Astronomical Society. **17**. 1976. pp263–289.

[64] Maurice Crosland. *Gay-Lussac Scientist and Bourgeois.* Cambridge University Press 1978.

[65] J R Mayer. *Bemerkung über die Kräfte der unbelebten Natur.* Justus Liebig's Annalen der Chemie. Vol. **42**, Issue 2, 1842. pp233–240.

[66] H E Roscoe. *The Edinburgh Review: or Critical Journal.* Vol cxix p1. January 1864.

[67] J Smeaton. *An experimental enquiry concerning the natural powers of water and wind to turn mills, and other machines, depending on a circular motion.* Phil. Trans. **51**. January 1759. pp100-174.
J Smeaton. *An experimental examination of the quantity and proportion of mechanic power necessary to be employed in giving different degrees of velocity to heavy bodies.* Phil Trans. **66** January 1776. pp450-475.

[68] P Ewart. *On the measure of moving force.* Memoirs of the Lit. and Phil., 2nd ser., Vol. **2**. April 1829. pp105–258.
P Ewart. *Experiments and observations on some of the phænomena attending the sudden expansion of compressed elastic fluids.* Phil Mag. *January 1829, pp247–254, being written extracts from two papers read before the Manchester Literary and Philosophical Society.*

[69] Rev Reid. *An Essay on Quantity; Occasioned by Reading a Treatise, in Which Simple and Compound Ratio's are Applied to Virtue and Merit.* Phil. Trans. 45. vol. 45 pp 505-520 1 January 1748. For reasons I am unable to determine, Reid inserted an apostrophe in the word ratios, or else the editor of the journal did it for him, or it was common at the time.

[70] Rumford, Benjamin Count Rumford. *An experimental enquiry concerning the source of the heat which is excited by friction.* Phil. Trans. R. Soc. Lond. **88**. 1798 pp80–102. (doi:10.1098/rstl.1798.0006)
An Enquiry concerning the nature of heat, and the mode of its communication. Phil. Trans. R. Soc. Lond. **94**. 1804. pp77–182.

[71] John Davy. *The collected works of Sir Humphrey Davy, edited by his brother.* Vol II. London 1839. pp5–32.

[72] Adair Crawford. *Experiments and observations on animal heat and the inflammation of combustible bodies.* London. Murray and Sewell. 1788.

[73] Ainé Seguin. *Observations on the effects of heat and of motion.* Edinburgh Philosophical Journal. Vol X. 1824.

[74] Ainé Seguin. *De l'influence des chemins de fer.* Liège 1839.

[75] Ainé Seguin. *History of the dynamical theory of heat.* Proc. Lit. and Phil. Manchester Vol **III** 1864. pp21–24.

[76] *Manchester New College - Introductory Lectures.* Delivered at the opening of the session in 1840. Simpkin, Marshall and Co. and J Green, London 1841.

[77] *The Literary Gazette and the Journal of the Belles Lettres, Arts, Sciences, etc.* No 1158. London. Saturday, March 30, 1839.

[78] Stephens, Michael D. and Roderick, Gordon W. (1972) *Nineteenth century ventures in Liverpool's scientific education.* Annals of Science, **28**. 1, pp61–68.

[79] Montagu Lyon Phillips. *Worlds beyond the Earth.* Richard Bentley, London 1855. vii + 274 pages.

163

[80] William Miller. *The Heavenly Bodies. Their Nature and Habitability.* Hodder and Stoughton, London 1883.

[81] Frederick William Faber. *The Blessed sacrament; or The Works and Ways of God.* John Murphy & Co., Baltimore 1855.

[82] Matthew Mercer. *Dissenting academies and the education of the laity, 1750–1850.* History of Education, **30**. 1, 2001. pp35–38.

[83] J B Dancer. *Early Photography in Liverpool and Manchester.* Manchester City News, 22 May 1886.

[84] Henry Garnett. *Photographs of John Dalton.* Nature. February 7, 1931. p201.

[85] Henry Garnett. *John Benjamin Dancer, instrument maker and inventor.* Memoirs and Proceedings, Manchester Literary and Philosophical Society, LXXIII, Memoir 2, 1928. p7.

[86] A L Smyth FLA. Joint Honorary Librarian, Manchester Literary and Philosophical Society. *John Dalton 1766–1844 A Bibliography of works by and about him.* Manchester University Press. 1966.

[87] L L Arden. Memoirs and Proceedings of the Literary and Philosophical Society of Manchester. **104**, 1961–2, pp67–69.

[88] Editorial. *Palmam qui Meruit Ferat.* Photographic Journal, VI, no. 93, 1859 p104 and no. 94, 1859 p118.

[89] J R Ashworth. *Joule's thermometers in the possession of the Literary and Philosophical Society.* Journal of Scientific Instruments, Vol. **7**. 1930. pp361–363.

[90] Supplement to the Manchester Examiner and Times. Saturday 15th March 1851.

[91] Archibald Sandeman. *Pelicotetics or the Science of Quantity.* Deighton Bell & Co. London 1868.

[92] Charles Dickens (ed). *The Household Narrative of Current Events, (For The Year 1851,) conducted by Charles Dickens.* London 1851.

[93] *Essays and Addresses by Professors and Lecturers of the Owens College, Manchester.* MacMillan. London 1874.

[94] Ordnance Survey county series [cartographic material]: Lancashire first edition. Six-inch full sheet, Manchester City Centre. 1845. Southampton: Ordnance Survey.

[95] Slater's New Plan of Manchester and Salford. Isaac Slater, Manchester, 1871.

[96] Robert H Kargon. *Science in Victorian Manchester.* Manchester: The University Press 1977, The John Hopkins University Press 1977.

[97] J J Thomson. *Recollections and Reflections.* London: Bell 1936.

[98] A Schuster. *The Influence of Mathematics on the Progress of Physics. Introductory address delivered at the Owens College.* Manchester: J. E. Cornish, 1881; also Nature Vol. 25. p.p. 397-401. (1882).

[99] Ernst Von Meyer. *A History Of Chemistry.* 3rd English edition, translated from the 3rd German edition by George McGowan. MacMillan, London 1906.

[100] Robert Fox and Graeme Gooday (ed). *Physics in Oxford, 1839-1939: Laboratories, learning and college life.* Oxford University Press. 2005. pp80–118.

[101] *Obituary Notices of Fellows Deceased.* Proc Royal Soc. Series A, Vol. **99**. No 701 (Sep 1, 1921) pp vi-viii.

[102] Letter from O F Brown to A J Croft, 18 September 1968, Clarendon Laboratory Archives.

[103] William Jack. *Galileo and the application of mathematics to physics.* James Maclehose, Glasgow, publisher to the University. Macmillan. 1879.

[104] A W Ward. *Obituary of William Jack.* In The Sex, Magazine of the Peterhouse Sexcentenary Club. Cambridge: Palmer 1924. p15.

[105] William Jack. *On the galvanometer.* Proc. Lit. and Phil. of Manchester. Vol **VI** 1867. pp147–150.
Further Remarks on the galvanometer. Loc. cit. pp158–160.
On the galvanometer. Loc. cit. p177.

[106] Cargill Gilston Knott. *Life and scientific work of Peter Guthrie Tait.* Cambridge University Press. 1911.

[107] Osborne Reynolds. *On an Electrical Corona resembling the Solar Corona.* Proc. Lit. and Phil. of Manchester. Vol **XI** 1871–72. pp100–106.

[108] Osborne Reynolds. *On an inversion of ideas as to the structure of the universe.* Cambridge University Press. 1903.

[109] Osborne Reynolds. *The Sub-mechanics of the Universe.* Cambridge University Press. 1903.

Index

SDA Electronics Ltd

Units 30, 35 & 36, Willan Industrial Estate, Vere Street,
Eccles New Road, Salford M5 2GR

Telephone: 0161-745 7029 (Unit 30): 0161-736 9585 (Unit 36)
Fax: 0161-745 9649

Hi Andy, Review copy.

Alan.

With Compliments

21050333R00105

Printed in Poland
by Amazon Fulfillment
Poland Sp. z o.o., Wrocław